Springer Tracts in Electrical and Electronics Engineering

Springer Tracts in Electrical and Electronics Engineering (STEEE) publishes the latest developments in Electrical and Electronics Engineering - quickly, informally and with high quality. The intent is to cover all the main branches of electrical and electronics engineering, both theoretical and applied, including:

- Signal, Speech and Image Processing
- Speech and Audio Processing
- Image Processing
- Human-Machine Interfaces
- Digital and Analog Signal Processing
- Microwaves, RF Engineering and Optical Communications
- Electronics and Microelectronics, Instrumentation
- Electronic Circuits and Systems
- Embedded Systems
- Electronics Design and Verification
- Cyber-Physical Systems
- Electrical Power Engineering
- Power Electronics
- Photovoltaics
- Energy Grids and Networks
- Electrical Machines
- Control, Robotics, Automation
- Robotic Engineering
- Mechatronics
- Control and Systems Theory
- Automation
- Communications Engineering, Networks
- Wireless and Mobile Communication
- Internet of Things
- Computer Networks

Within the scope of the series are monographs, professional books or graduate textbooks, edited volumes as well as outstanding PhD theses and books purposely devoted to support education in electrical and electronics engineering at graduate and post-graduate levels.

Review Process

The proposal for each volume is reviewed by the main editor and/or the advisory board. The books of this series are reviewed in a single blind peer review process.

Ethics Statement for this series can be found in the Springer standard guidelines here https://www.springer.com/us/authors-editors/journal-author/journal-author-helpdesk/before-you-start/before-you-start/1330#c14214

Devika Jay · K. Shanti Swarup

Market Operation for Reactive Power Ancillary Service

Design and Analysis with GAMS Code

Springer

Devika Jay
Department of Electrical Engineering
Indian Institute of Technology Madras
Chennai, India

K. Shanti Swarup
Department of Electrical Engineering
Indian Institute of Technology Madras
Chennai, India

ISSN 2731-4200 ISSN 2731-4219 (electronic)
Springer Tracts in Electrical and Electronics Engineering
ISBN 978-981-99-6951-7 ISBN 978-981-99-6952-4 (eBook)
https://doi.org/10.1007/978-981-99-6952-4

This Springer imprint is published by the registered company Springer Nature Singapore Pte Ltd.
The registered company address is: 152 Beach Road, #21-01/04 Gateway East, Singapore 189721, Singapore

Paper in this product is recyclable.

Preface

This book on Market Operations for Reactive Power Ancillary Services is a book aimed for both graduate and postgraduate students of engineering colleges. Electric markets and its operations are gaining more importance after Deregulation and Restructuring of Electric Utilities. The unbundling of services and the participation of private partners in generation, transmission and distribution has further enabled the increased interest in various types of trading in electric markets. This book intends to provide information to the reader on the recent developments in the market operations for energy trading in the power sector. Ancillary services are an important aspect for trading in electricity markets. Ancillary services are usually ignored in most of the calculations and are taken for granted. Understanding ancillary services is an important contribution to this book. This book is divided into three parts as nine chapters and two appendices. The chapters cover the whole spectrum of Electric Markets and Energy Trading, importance of reactive power as an ancillary service. New methodologies and techniques like game theory and deep learning are introduced in the last part of the book.

The book is unique and important from different perspectives of academic and industry to familiarise and update themselves of the recent developments in energy trading in electricity markets. The purpose is to bring awareness to the students, research scholars, practising engineers and power engineering community to understand the present problems encountered in electric power systems. The book covers the complete spectrum audience of students, operating and maintenance personal and engineering college faculty on how to understand and solve problems in real-world situations. The reactive power ancillary services in the Indian context are described in a separate chapter. The book is expected to be useful in understanding, formulating and solving problems using the GAMS code provided in appendix B. GAMS is a widely useful software in solving problems in the power engineering discipline. Using the examples and the code provided the readers of the book can easily extend it to practical and realistic problems encountered.

The book is organised as three parts covering eight chapters and two appendices containing additional modules of GAMS code and reactive power voltage support. Part I deals with introduction to electricity markets and deregulation that are provided with emphasis on energy and ancillary services. Three chapters are provided in this part of the book. Reactive power ancillary service is an important aspect which is usually ignored in the current literature. Part II describes the design aspects of reactive power markets. The part highlights the need to consider the localised nature of reactive power through partitioning the grid into localised reactive power market. The part also contains an important chapter on the value-based market formulation which is relevant in the case of reactive power ancillary services. The final Part III discusses the analysis of reactive power markets with new developments like game theory and deep learning as two chapters. Part III also includes a review on the importance of reactive power ancillary service markets in the Indian grid. The insights provided in this chapter are based on the information and discussions with the GRID-INDIA (Grid Controller of India Limited), previously known as Power System Operation Corporation (POSOCO). The appendices are important and supporting contributions to the book. These are useful in understanding the concepts underlying the book and ease of implementing the ideas for better operation and planning of the power systems. The GAMS code helps in further academic research and also in industrial R&D activities in studying the impact of reactive power ancillary service in their grid operations.

Chennai, Tamil Nadu, India

Devika Jay
K. Shanti Swarup

Contents

About the Authors

Devika Jay completed her M.S. (Research) and Ph.D. in Electrical Engineering from IIT Madras in 2013 and 2021, respectively. She is currently the Chief Technical Officer/Chief Operating Officer of Gridsentry Private Limited, a substation cybersecurity startup incubated at IISc Bangalore and IIT Kanpur. Her research interests include deregulation, electricity markets, cybersecurity, game theory, and AI applications in power systems. She has many research papers in journals and conferences on smart power systems, game theory, artificial intelligence, and machine learning for power system applications. She has won several awards during her academic career. She served as deputy manager at Power Grid Corporation of India Limited (PGCIL) until July 2021, where she was in charge of the design, planning, and operation of/ 400/220 kV substations.

K. Shanti Swarup is a faculty with the Department of Electrical Engineering, Indian Institute of Technology, Madras, India. He completed Ph.D. from the Indian Institute of Science Bangalore in 1991. Before joining the department as a visiting faculty member, he held positions at the Mitsubishi Electric Corporation, Osaka, Japan, and Kitami Institute of Technology, Hokkaido, Japan, serving as a visiting research scientist and visiting professor, respectively, from 1992 to 1999. Since 2000, he has been a professor at IIT Madras. His research areas include Power Systems, Smart Grids, artificial intelligence, knowledge-based systems, computational intelligence, soft computing, Energy Management Systems (EMS), Supervisory Control and Data Acquisition (SCADA), Power System Automation, and Network Protection. He has done sponsored research projects with various industries like BHEL, Hitachi, Easun-MR, MR-GmbH, etc.

Part I
Deregulation in Power System

Chapter 1
Introduction to Electricity Markets

Abstract This chapter presents the key concepts associated with deregulated power systems and electricity markets. The basics of markets and power system economics are also discussed in the chapter.

Keywords Deregulation · Restructured power systems · Electricity markets · Power system economics

1.1 Deregulation in Power Systems

When electricity was comparatively a new technology, investment in electrical technology was regulated by Government to avoid monopoly and exploitation of end consumers. Hence generation, transmission and distribution were bundled as a single service regulated by a central utility in traditional power systems. This leads to a vertically integrated utility structure for power systems as shown in Fig. 1.1.

Under vertically integrated utility (VIU) structure of power systems, customers dealt with the central utility. Thus the power business had only two entities then: the monopolist utility (generation/transmission/distribution) and the customer.

The disadvantage of such vertically integrated utility structure of power systems is the power industries enjoyed monopoly in their area of operations. To check the prices set by monopolist entities regulations are imposed by imposed by Government or Government authorities. Regulation is a set of rules that define a framework for smooth and disciplined power flow ensuring that there is no undue advantage to any particular entity.

Thus, under VIU structure, the work nature of utilities is defined by the regulatory framework setup by the Government. Utility should meet the demand from customers, and the return on utility's investments is regulated by the Government. Utility should provide power to all those customers who demand for it, and the return on the utility's investment is regulated by the Government.

However, the monopoly status of electric utilities will not provide any incentive for efficient operation. To improve the investment in power sector for improving the electricity infrastructure in the country and to adopt newer technologies that

D. Jay and K. Shanti Swarup, *Market Operation for Reactive Power Ancillary Service*,
Springer Tracts in Electrical and Electronics Engineering,
https://doi.org/10.1007/978-981-99-6952-4_1

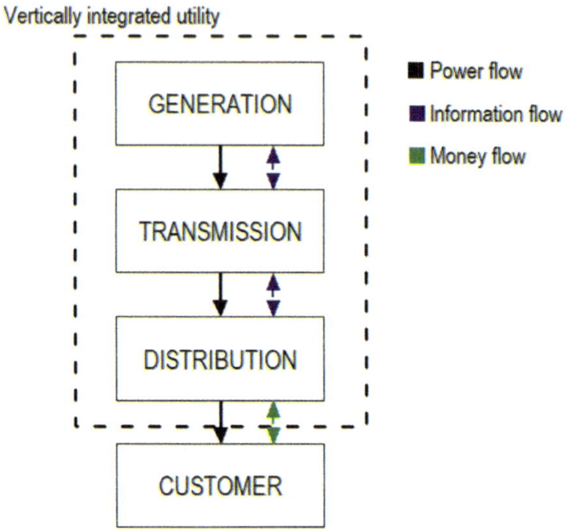

Fig. 1.1 Vertically integrated utility structure of power systems

will ensure quality and reliability in power supply, monopolistic nature of vertically integrated utility structure was found to be inefficient. Economists are of the opinion that by inducing competition in electricity sector would improve participation of more private firms. This will help in encouraging efficient operation of the utilities and investments in power infrastructure to ensure power supply quality. Such competitive environment was found to be successful in sectors like airline, gas, telephone, etc. Competitive environment helps in bringing down the cost of electricity, and power industry activities become more customer centric.

These observations led to the concept of deregulation in power sector. This is also called as restructuring of power industry which means discontinuation of the framework provided by the regulation. Through deregulations, prices of various services in power sector are determined through a competitive market environment. The commonly used word in Europe is 'liberalisation' of power industry; 'deregulation' is a more popular phrase in USA. Deregulation in power industry implies restructuring of the set of rules set up by the Government to control power industry.

With the advancement of electricity technology, a new dimension status as a 'commodity' was provided for electrical energy. This makes electrical energy 'commodity' suitable for trading, i.e. for buying and selling, in a market place under a competitive environment. This helps in increasing the participation of private players in electricity sector. The advantages of such competitive environment are

1. Electricity prices will be less than monopolistic prices.
2. Consumers can choose better retail utilities that provide then better price, service and high-quality reliable power supply.
3. Upon receiving better payments, private players increase investment in innovation to improve service and technology to survive in the market.

Deregulation of power sector first involves identification and separation of essential services from the bundled VIU structure. This process is called unbundling. This segregation is necessary to identify those services which can be treated as commodities and traded in a market place. Certain services may still be treated under regulatory bodies or Government authorities. The difference between unbundled systems and VIU system is that in VIU all services are handled under a single umbrella through regulations set by Government. In unbundled systems certain services are separated and treated as commodities for buying and selling with prices determined through competitive environment.

The restructuring process of power sector is not uniform in all countries. However, the general concept of deregulation/restructuring in power systems involves splitting up of the vertically integrated utility into unbundled companies.

1.1.1 Entities in Restructured Power Systems

The process of deregulation or restructuring in power systems started with the separation of transmission services from generation activities. This led to the formation of generation companies (GENCOs). Further upon to improve the quality of service provided directly to the consumers, distribution services can be separated from transmission to form two more entities: transmission companies (TRANSCOs) and distribution companies (DISCOs). Thus, competition persists in generation, transmission, distribution as well retail energy supply.

The functioning of unbundled entities, namely GENCOs, TRANSCOs and DISCOs, is regulated and monitored by an Independent System Operator (ISO). The structure of deregulated power systems is as shown in Fig. 1.2. We shall discuss the function of each entity shown in Fig. 1.2 in detail:

1. GENCO: Generation company (GENCO) comprises one or more generators that sell power into competitive marketplace.
2. TRANSCO: Transmission company (TRANSCO) owns and maintains the power transmission facilities and involves in the management of assets to improve power quality and reliability. Thus TRANSCO is responsible for bulk transmission of power from GENCOs to consumers through distribution sector.
3. DISCO: Distribution company (DISCO) operator is local power delivery system to end-users like industrial, commercial and residential consumers.
4. ISO: Independent System Operator is third-party entity which ensures reliable power supply in the system. This third party is required to be an independent authority and does not participate in energy trading. This is required to ensure there is no conflict of interests. The prime responsibility of the ISO is to ensure a safe and secure operation of the grid by properly coordinating, controlling and monitoring the major entities of the deregulated systems, i.e. GENCO, TRANSCO, DISCO and end-users. The intention of the ISO is expected to be neutral, and transparency in its actions is to be ensured. The basic functions of ISO are

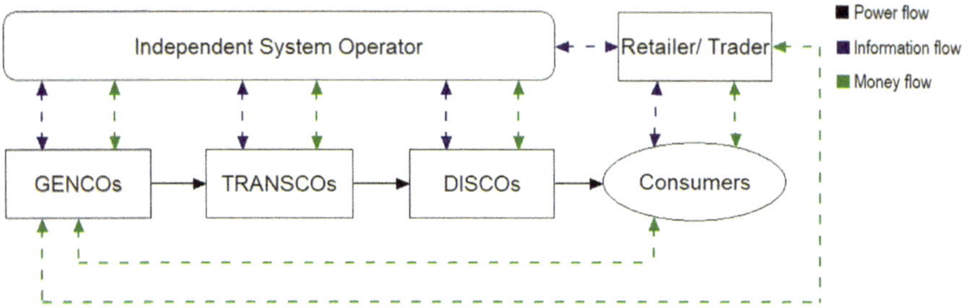

Fig. 1.2 Restructured power systems

a. Ensure stable grid operation.
b. Ensure availability of transmission assets for buying and selling of power.
c. Determine suitable pricing mechanism for services unbundled as commodities traded in marketplace.
d. Ensure power quality and reliability.
e. Promote economically efficient operation in the system.

5. Retailer/Trader: These entities buy power from GENCOs to sell it directly to consumers at a competitive retail price. It is to be noted that these entities do not own any network physical assets.

1.1.2 Challenges in Deregulation

Though deregulation in sectors like airline, gas, etc. is found to be successful, deregulation process in electricity sector is challenging. This is because, treating electricity as a commodity is not straightforward considering the dynamics and operation limitations in the electricity network. These operation constraints impose several challenges in the deregulation/unbundling process of power systems as discussed below:

1. Handling network congestion: Any arbitrary transaction, i.e. buying and selling of power cannot be approved by the ISO. This is because each transmission network has operating constraints like line flow limits, bus voltage magnitude limits, etc. ISO must ensure that the power transaction contracts between buyers and sellers should be in such a way that these constraints are preserved to avoid adverse affects in the network like system blackout. In regulated system the sole responsibility of the VIU is to maintain transmission network constraints.
2. Provision of ancillary services: Activities that are necessary to support power transmission and at the same time ensure safe and secure operation of the grid. These services include scheduling and dispatch, frequency regulation, voltage control, generation reserves, etc. and are commonly referred to as ancillary ser-

vices. In deregulated power systems, ancillary services are unbundled and priced separately. System operators may have to purchase ancillary services from ancillary service providers.
3. Design of market mechanisms: Defining suitable market mechanisms includes consideration of dispatch criteria, pricing schemes, definition of trading window. Also separate mechanisms are to be defined for various ancillary services.
4. Market power and price volatility: Certain services in the network may result in exercising of monopoly by private players. Such situations arise due to intentional or unintentional utilisation of opportunity created by network operating constraints. This may result in undesired hike in prices which needs to be investigated and rectified through proper investment in infrastructure. Prices are to be kept under control through devising suitable regulatory measures and policies.

1.2 Electricity Markets Under Deregulation

Under deregulation, energy and ancillary services are treated as commodities which involve buying and selling by market players in market place. This has resulted in the formation of electricity markets in restructured power systems. However, during these transactions ISO must ensure the stable grid operations with system parameters within the permissible limits. Hence the role of ISO being independent of market players is very much significant in electricity market operations.

Considering the significance of market operation in restructured power systems, the Federal Energy Regulatory Commission (FERC) Order No. 888 mandated the establishment of unbundled electricity markets. Energy and ancillary services are offered as unbundled services. Generating companies (GENCOs) compete to sell energy to customers through electricity markets by submitting competitive bids. This would enable GENCOs to maximise their profits and thereby encourage them to acquire and invest in computational tools like load forecasting, price forecasting, unit commitment, etc. to improve the status in the competitive market environment.

The primary objective of market operation is to ensure safe and secure operation of the grid. Under deregulation, security is achieved by utilising diverse services through market operations available. Another advantage of market operation is that it enables economic operation of the system. Competitive markets enable to reduce cost of electricity utilisation.

1.2.1 Modes of Market Operation

There are various modes of market operation based on the hand-shaking timeline between buyers and sellers. Generally market operations can be of two modes

1. Bilateral markets.
2. Spot markets.
   - Day-ahead market.
   - Real-time market.

In forward markets, buyers and sellers trade through non-standard, long-term, bilateral contracts through mutually agreed price. Spot market includes day-ahead and real-time market. A customer who buys power in a forward market will receive electricity delivered by the seller. If not received, customers are entitled for a financial compensation. An agreed transaction through day-ahead market may in some case not be delivered in real-time. This results in real-time markets for separate real-time price of energy.

A way to distinguish between forward and spot markets is by considering day-ahead and real-time markets as spot markets, while all trades taking place before that are termed as forward or bilateral trades. In bilateral trades, there is no single market price as such. In the rest of the module, we prefer to define the spot market as defined just above. Much ahead of real-time, i.e. more than a week, month or years ahead, bilateral contracts provide the best manner of trading power. Settlements of bilateral contracts take place very slowly.

1.2.2 Models for Market Operation

There are mainly three models for market operation, namely

1. PoolCo model.
2. Bilateral contract model.
3. Hybrid model.

In PoolCo model, sellers and buyers submit bids to the pool for the amounts of power that they are willing to trade in the market. PoolCo model is a centralised marketplace for clearing market. Sellers supply energy to the grid and not to specific customers. Sellers with very high bids are not allowed to sell. Similarly buyers with too low bids are not allowed to buy. Thus low-cost GENCOs are encouraged in PoolCo market. Economic dispatch is performed by ISO to determine a single (spot) price for electricity. The competitive level of the spot price is equal to the marginal cost of the most efficient bidder, and this is driven by market dynamics.

Negotiable agreements are arrived on delivery of power between two market players in a bilateral contracts model. The contact agreements and rates are decided independent of ISO. The responsibility of ISO is to clear the actual transactions by verifying the available transmission capacity and other system constraints to preserve the security of the grid. The advantage of bilateral contracts model is the flexibility offered to trading parties in determining prices and contract terms. However, there is risk on credibility of counterparts and high cost of negotiating and writing contracts.

The hybrid model combines the advantages of PoolCo and bilateral contract models. PoolCo model is availed by participants who do not sign bilateral contracts. Through bilateral contracts, customers get an opportunity to negotiate and arrive at Power Purchase Agreements that will suit their needs. This will help in improving the quality of service and reliability.

1.2.3 Types of Electricity Markets

Based on the commodity for trade, electricity markets can be classified into the following types:

1. Energy markets.
2. Ancillary service markets.
3. Transmission service markets.

A market mechanism that facilitates energy trading between buyers and sellers is called energy markets. The clearing and settlement function in energy market is neutral and independent. ISO or power exchanges operate the energy market.

Ancillary service markets are significant because these services are needed for the reliable operation of power systems. Under deregulation, these services are unbundled from the energy supply service and hence procured separately through competitive markets.

In transmission service markets, the right to transfer power from the point of injection to the network or from the point of withdrawl from the network is traded. Thus in transmission service markets, transmission right is the commodity. A centralised market mechanism is conducted by ISO or party appointed by ISO to collect bids for purchase and sale of transmission rights. The objective of the market is to clear the bids such that the transmission constraints are considered and ensure economic operation of transmission network.

1.3 Concepts of Smart Grid

With the advent of advanced information and communication technology (ICT), integration of small-scale distributed energy sources, etc., technological advancements in power system operation and control were also implemented. Thus generation, transmission, distribution and consumers were linked with an improved and advanced information and communication technology leading to fast and reliable grid operations. These features led to the concept of smart grid, where in system operator utilised the advanced communication technology for system operation and control. National Institute of Standards and Technology (NIST) USA defines smart grid as

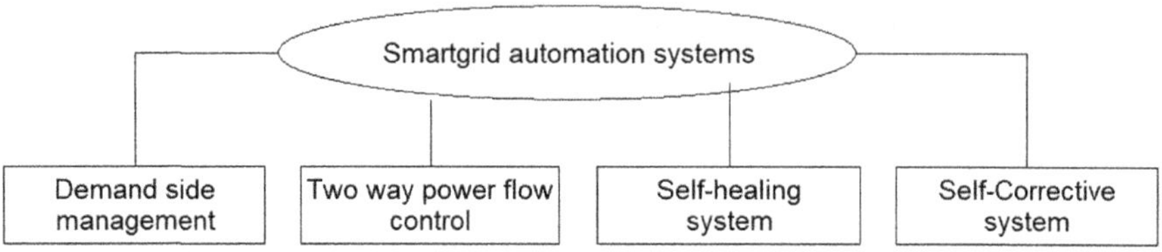

Fig. 1.3 Essential features of smart grid

> A modernised grid that enables bidirectional flows of energy and uses two-way communication and control capabilities that will lead to an array of new functionalities and applications.

As per IEEE, a large 'System of Systems' with three functional layers, namely: power and energy layer, communication layer and IT/Computer layer, in which communication and IT layer make the power and energy layer smarter.

The main objectives of a integrating ICT with power systems are

1. Facilitate Demand Side Management: Provide customers with better prices and allow them to make usage decisions that will enable efficient energy management system.
2. Facilitate System Automation: Through system automation techniques, reliability of the system can be improved.
3. Integration of Renewables: Better controllability of low inertia renewable energy sources can be obtained with advanced ICT, which will help in integration of renewables to control operations much more easier.

Smart grid is featured with distributed energy resources which are located closer to the grid. This helps in serving customers better, and also customers are allowed to choose their supplier. The advantage of such distributed sources is that investment in transmission and distribution is reduced. It also reduced energy losses, risks of major outages, improves reliability, etc. With improved sensors, remote monitoring, automated switches, etc., end-to-end monitoring and control of power flow is possible. This is possible with the implementation of advanced ICT into the network leading to a smart grid. Similarly, end-users are equipped with advanced metering infrastructure, automated home control systems, smart appliances, etc. which will enable customers to suitable manage their energy consumption without losses and at the same time without compromising their comfort utility. Near real-time meter data will enable customers to understand energy usage and consumption of individual smart appliances. Consumers may also be equipped with renewable energy sources and energy storage systems which elevate their status from consumers to prosumers.

The key functions of a smart grid achieved through ICT are as shown in Fig. 1.3.

The co-existence of various types of renewable and non-renewable generating technologies is a salient feature of smart grid. Coordination of these varied resources of power shall be achieved by system operator using communication and information infrastructure through Wide-Areas Measurement System (WAMS). Thus for making the grid smarter from generation perspective, communication infrastructure is required to be placed at generation units for efficient system operation and market environment.

In smart grids, a communication interface exists between transmission network, generating station, system operator, market operator and distribution system. This helps in monitoring the transmission network in real-time and protects the system from potential contingencies and blackouts. Power flow and voltage monitoring on lines are easily controlled using the communication interface between sensor like Phasor Measurement Units. This enables the system operator to utilise the network optimally to reduce losses and voltage deviation in the network and maximise supply reliability.

Smart distribution systems are enabled by substation automation, distribution automation techniques and distributed energy resources (DERs). System operation at smart distribution level is carried out by independent distribution system operator. Hence information exchange between DERs, prosumers (consumers with storage or independent generation) and microgrids is essential for smart distribution system.

Another key feature of smart grid is the capability of consumers to shape their usage according to price signals issued to them, i.e. demand response. Home (building) automation systems help consumers to control and monitor their power consumption from smart appliances and thereby schedule their usage according to peak and offpeak load timings. This can be achieved through a proper communication infrastructure which will help consumers to interact with system operators (distribution) and electricity market.

Smart grid infrastructure leads to emergence of new class of entity in grid operations called service providers. Service providers are required to forecast renewable generation to handle uncertainty associated with such sources. Since most of the essential grid services are achieved through market operations billing and customer management service providers, accounts management, home automation system providers, energy storage systems will also be required for smooth functioning of smart grid.

Electricity markets are to designed in smart grids considering the aforementioned scope of generation, transmission, distribution and consumers. The real-time pricing signals are to be made available at the end-user for facilitating demand response and demand side management by system operator. Better prices are to be issued and made available in real-time to encourage customers to wisely use their appliances and also utilise renewable energy sources for green and clean grid operation.

Thus, two types of markets are to be studied in the context of smart grids—wholesale markets and retail markets. Wholesale markets are concerned with the electricity markets pertaining to transmission level. Retails markets are required to coordinate distributed energy sources, prosumers, microgrids, etc. at the distribution level. These two levels of markets are also required to be well coordinated to improve the smartness of the grid.

1.4 Wholesale Electricity Markets

When electricity is traded before being delivered to end-customers through the network, the market place is called wholesale electricity markets. The fundamental components in wholesale electricity supply are generation, transmission and coordination services before being delivered to the end-users. Hence market participants in wholesale electricity markets are GENCOs, TRANSCOs, traders and demand side management facilitators. The primary objective of wholesale market is to ensure constant balance of generation and load. The uncertainty in generation and transmission must also be coordinated well through wholesale markets to ensure reliable supply to consumers.

The services that are traded through wholesale electricity markets are

1. Scheduled energy through long-term contracts, day-ahead, intra-day and balancing markets.
2. Ancillary services like frequency control (primary, secondary, tertiary reserves), reactive power for voltage regulation and black start restoration.
3. Generation capacity.

There are various ways in which wholesale electricity markets are operated. Major classification for the mode of operation in wholesale market is spot markets and forward markets. These markets are majorly characterised by bilateral contracts through physical and financial contracts. Operation of spot markets is summarised in Fig. 1.4.

Spot markets are typically day-ahead markets, and the spot market prices are reference to other transactions. However such markets are subject to arbitraging between other markets. Arbitraging implies the taking advantage of price difference between two or more markets. In addition to day-ahead markets, intra-day markets are also required to handle contingencies like unexpected outage of generation/load and to balance supply and demand. These are additional transactions and do not modify the existing and settlements cleared in day-ahead and other short-term and long-term markets.

In forward contracts, buyer is committed to purchase and the seller is committed to deliver the specified quantity in specified time period (future) at a price predetermined through contracts. Forward markets are operated through bilateral contracts

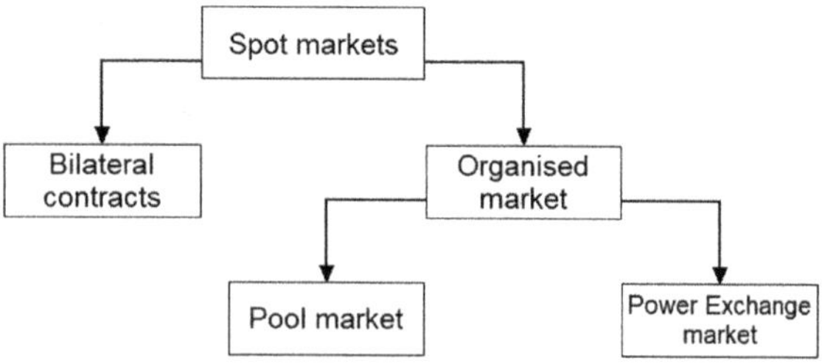

Fig. 1.4 Functioning of wholesale spot markets

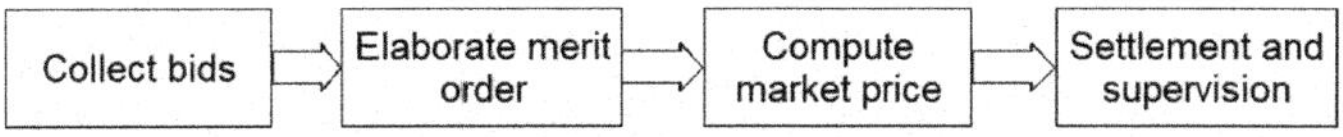

Fig. 1.5 Role of power exchanges in wholesale market

or organised markets which are operated through physical and financial contracts. Financial contracts involve only cash delivery on expiry. Exchange or pool prices are considered as reference for financial contracts.

The features of physical contracts are

1. Involve physical and cash delivery on expire.
2. The delivery point is a node on the high-voltage grid.
3. The right to transport power has to be attained by participants.
4. The transactions are to be approved by the system operator to maintain network constraints within the bounds.
5. Mechanism is required at transmission system operator level to settle and manage real-time imbalance.

Power exchanges play a significant role in the implementation of wholesale electricity markets. The functions of power exchanges are shown in Fig. 1.5.

Participation of power exchanges in wholesale markets can be either mandatory or voluntary. Significance of power exchanges is prominent in day-ahead markets. To improve the suboptimal results and modify the commitments in day-ahead markets, power exchanges provide adjustment markets. The system contingencies like unplanned outages may also be considered in the adjustment markets.

In the USA, wholesale markets are based on locational marginal pricing. The settlement period may be day-ahead or real-time (5-min time block). Day-ahead markets are based on complex bids and unit commitment algorithms. Real-time markets are based on optimal power flow. Long-term markets are also envisaged for transmission network planning at ISO level. New England ISO and PJM have an auction-based forward capacity market carried out once in 4 years. In New York a demand curve-based monthly capacity market is carried out.

1.5 Retail Electricity Markets

With the emergence of distributed energy resources like distributed generation, demand response compatible loads, energy storage devices, etc., and its high penetration in smart grids, wholesale markets alone are found to be not sufficient for management of distribution grid. Hence distributed market mechanisms are required instead of a centralised perspective as in wholesale market. Such mechanism will help in formulating efficient scheduling and compensation. Distributed market mechanism is based on local computations at each node which is computationally efficient.

Peer-to-peer exchange of information among entities of the distribution grid is also essential for efficient scheduling of distributed energy resources. However design of such distributed markets is resilient to communicated link failure, noise and delays.

This concept has led to the design and implementation of retail electricity markets. Distribution system operator functions as the third-party independent authority that is capable of handling retail electricity markets. In retail electricity markets, end-users can choose their supplier and face real-time pricing. Dynamic pricing is an essential feature of retail market. Time of use price, critical peak pricing and real-time pricing shall be issued in retail markets to facilitate demand response.

An advantage of retail electricity market is that it encourages innovation among suppliers to provide green and reliable supply at better prices. However chances for irregular practices are possible in retail markets due to the asymmetry access to information from all retailers to consumers.

1.6 Coordination of Wholesale and Retail Markets

Wholesale markets are useful for achieving supply–demand balance through price signals like locational marginal pricing (LMP) obtained by solving optimal power flow problem at transmission level. However, retail markets are essential to integrate the participation of small-scale consumers in the distribution level. To exploit the capabilities of demand response and demand side management in full, dynamic pricing schemes in retail markets are to be envisaged. This has been achieved through distribution LMP (DLMP), which is determined by the distribution system operator using AC-OPF at distribution level. The components of DLMP include active power price, reactive power price, congestion component, voltage component and line loss. However for effective operation of the grid, the calculation of DLMP cannot be employed independently and locally. This is because the response of demand to DLMP will have a global influence in power system operation. Hence coordination of wholesale market with retail market is essential.

Large-scale consumers participate in wholesale spot energy markets. Small-scale consumers are involved in retail markets, and most of the retail markets are characterised by fixed prices. The prices in retail markets are generally decoupled from the wholesale market prices, and thus the retail market prices do not reflect the power system dynamics. The disadvantage of decoupled wholesale and retail market operation is that demand side management becomes passive and the load controllability feature cannot be utilised in power system dynamics. Through proper coordination of wholesale market pricing and retail market pricing, demand response schemes can be designed to suit and facilitate the safe and secure operation of transmission and distribution network.

The general concept of coordinated wholesale and retail markets is depicted in Fig. 1.6. There are mainly three different coordination schemes between wholesale and retail electricity markets:

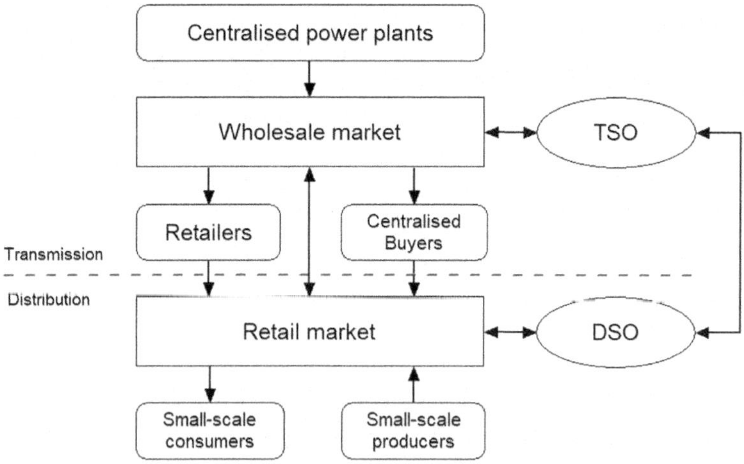

Fig. 1.6 Coordination of wholesale market with retail market

1. DSO leader coordination scheme.
2. DSO follower coordination scheme.
3. TSO-DSO iteration scheme.

These schemes have been defined on the basis of leader–follower models in which coordination is achieved between wholesale (TSO) and retail markets (DSO). The market operation which is considered in priority is modelled as leader where as the market which follows the first-move market is modelled as follower.

1.6.1 DSO Leader

In DSO leader-based coordination schemes, DSO makes the first move by clearing the retail market operation. If demand is still not satisfied or if there exists excess supply, DSO is capable of importing or exporting from the high-voltage transmission grid through wholesale markets from TSO. Thus, DSO has the priority over the TSO and allocated resources from distribution grid. On solving the local grid (distribution grid) constraints, DSO aggregates and offers the remaining bids to the wholesale market operated by TSO.

Some distribution companies may be proactive and described as proactive DISCO (PDISCO). These are capable of managing distributed resources and coordinating with the DSO to satisfy the network operating constraints. Such PDISCO submits continuous offers and strategical bids to the wholesale market. Such PDISCOs can be modelled as a distribution generation with the objective of profit maximising. In this case PDISCO is an aggregator that offers their best aggregated offer based on the bids received from distributed generators to the wholesale market.

A smart energy service provider (SESP) can be envisaged as a communication platform to facilitate the trading and scheduling of energy and other services in retail market. The objective of SESP is to maximise the social welfare for its participants under the retail market. SESP shall be capable of performing as an aggregator to participate in wholesale markets that will supplement the retail market operation. Retail markets thus will be capable of supporting power system balancing in the wholesale market.

1.6.2 DSO Follower

In this coordination scheme, wholesale market is considered as the first mover (leader) which will fix the prices and quantity of power to be transacted between transmission and distribution grids. Based on the transaction clearance approved by the wholesale market, retail markets dispatch their local resources.

Thus in DSO follower-based coordination scheme, the retail market operator maximises the social welfare of its resources based on the bids and the wholesale day-ahead prices and quantum of power cleared through market transactions. Thus the injection to the distribution network is fixed and decided by the wholesale imbalance clearing. This is a disadvantage of DSO follower scheme. Such fixed injections/withdrawl between distribution and transmission grid may result in non-optimal utilisation of local resources.

1.6.3 TSO-DSO Iteration

Coordination between TSO-DSO can be achieved by modelling the system as multi-level hierarchical coordination mechanism. This is achieved by iteratively communicating a generalised bid function (GBF) from retail market (DSO) to wholesale market (TSO). The generalised bid function is formed from the AC-OPF of the distribution network. The objective is to minimise the dispatch cost of the distributed energy resources in the retail market. The advantage is that the GBF derives a linear relation for the dispatch cost in the distribution network and power transfer between the distribution and transmission network. A residual supply function can also be considered to achieve a common TSO-DSO market model, which is passed from DSO to TSO to solve the power balance constraints.

The different coordination schemes are summarised in Table 1.1.

Table 1.1 Coordination between wholesale and retail markets

Coordination scheme	Advantage	Disadvantage
DSA leader	Efficient allocation of resources through social welfare maximisation	Market power in DSO
DSA follower	Operation is easy	Distributed energy resources are not utilised optimally
TSO-DSO iteration	Social welfare maximisation of whole system and optimal utilisation of DERs	Computational cost is high

1.7 Basics of Market Economics

Every market comprises two key players: producer and consumer. Hence the study of market economics involves modelling of these two players to determine the conditions of market equilibrium and market clearing prices.

1.7.1 Producer Model

A producer tends to produce more goods/commodity when the prices are high to earn more profit. However, the producer may also refuse to produce if the prices are below the threshold value determined by the choices of the producer. This threshold price is called opportunity cost. Thus the supply function of a producer is as shown in Fig. 1.7. In Fig. 1.7, π is the market price and 'Q' is the quantity produced at the market price. For a producer with opportunity cost equal to the market price π, the quantity produced is called marginal production. Any price greater than π will ensure an increase in production of the marginal producer. This is called extramarginal production. If price is less than π, the marginal producer will incur loss, and this quantity is called inframarginal production.

Elasticity of supply curve is defined as

$$\epsilon_s = \frac{\mathrm{d}Q/Q}{\mathrm{d}\pi/\pi} \tag{1.1}$$

The elasticity of supply (ϵ_s) is always positive. This will be higher in the long run than in the short run because suppliers have more opportunities to increase the means of production.

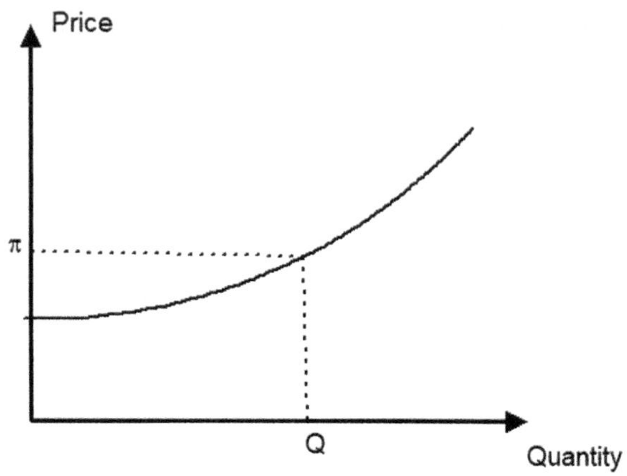

Fig. 1.7 Supply function—producer model

Fig. 1.8 Demand function—consumer model

1.7.2 Consumer Model

Consumer models are defined by the law of diminishing marginal utility, i.e. the amount consumed decreases as the price increases. The demand curve is as shown in Fig. 1.8.

The demand curve gives the marginal value that the consumer attaches to the commodity. It can be noted that the marginal willingness to pay decreases as the consumption increases. Hence the consumer model follows the law of diminishing marginal utility.

The elasticity of demand curve is given by

$$\epsilon_{\mathrm{d}} = \frac{\mathrm{d}q/q}{\mathrm{d}\pi/\pi} \tag{1.2}$$

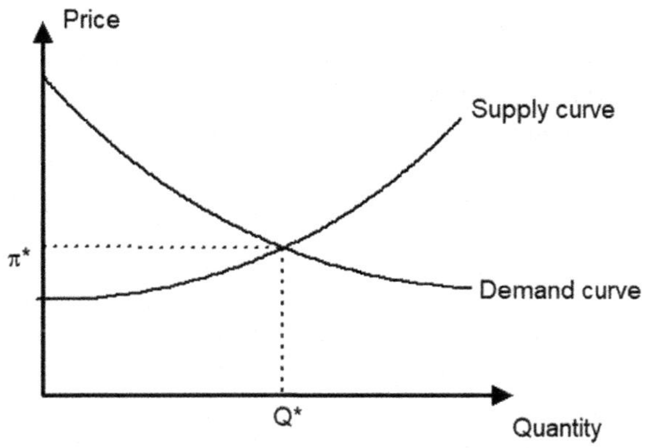

Fig. 1.9 Market equilibrium

For a given percentage change in price, if the change in percentage demand is high, then the demand is said to be elastic. If the change in percentage demand is less, demand is said to be inelastic.

1.7.3 Market Equilibrium

To determine the market equilibrium, we need to consider how producers and consumer interact. To simplify the analysis we consider a perfect competition scenario in which participants bid their actual cost. Also under perfect competition, a single player is not capable of changing the market price, and all buyers buy the commodity at the same market clearing price. The market equilibrium is achieved at that point at which quantity that the suppliers are willing to sell is equal to quantity that the consumers wish to obtain. The price that corresponds to this point is called market clearing price. Thus market equilibrium can be defined as the state in which there is zero excess demand and zero excess supply.

Figure 1.9 represents the interaction between producer and consumer to determine the market equilibrium which is the state with zero excess demand and supply. This point is achieved at the quantity 'Q^*' as shown in Fig. 1.9. The price that corresponds to this quantity is 'π^*' and is called the market clearing price. For a price less than 'π^*', the demand is more than the supply quantity willing to be generated by the producers. This excess demand will create more competition resulting in an increase in price. Similarly for a price greater than 'π^*', the supply is more than the demand quantity willing to be consumed at the customer level. This excess supply will reduce the price in the market. Equilibrium in quantity is achieved at the market clearing price 'π^*', and this defines the market equilibrium point.

In a competitive market, equilibrium is Pareto efficient. This implies that the benefit of any one party at the equilibrium point can be improved only be sacrificing the benefits of the other party. If total amount traded is less than the equilibrium

Q^*, the situation is not Pareto efficient. Similarly, any quantity greater than Q^* results in an price lower than the price it would take to supply if someone would be willing to pay for an extra unit. Hence this is also not a Pareto efficient situation. Thus Pareto efficiency is achieved when commodity allocation occurs on the basis of single marginal rate of substitution in a competitive market.

1.8 Summary and Takeaways

This chapter introduced the concept of deregulation and how traditional power systems operation differ under deregulation. The mode of procurement of services in deregulated power systems is through buying and selling of services in a marketplace. Thus electricity markets play a crucial role in maintaining the system parameters within the permissible limits and thereby ensure safe and secure operation of the grid. Hence the chapter discussed a brief background on the general mode and types of electricity markets. The basics of economics required to understand the concept of electricity markets were also discussed in this chapter which will serve as a prerequisite in understanding the concepts that will be discussed in chapters ahead. To illustrate the market equilibrium principle in electricity markets, energy markets and definition of locational marginal prices (LMP) were also presented in this chapter.

Chapter 2
Energy and Ancillary Service Markets

Abstract This chapter presents the generalised concepts of smart grid and how market operations are significant in a smart grid. The role of wholesale markets and retail markets is prominent in smart grids, and the necessity for coordinating these two markets is discussed in this chapter.

Keywords Energy markets · AC optimal power flow · Locational marginal price · Ancillary service markets

2.1 Energy Markets

In electricity markets, market equilibrium is achieved mostly through mechanism based on locational marginal price (LMP). The LMP mechanism was first invented by Dr. William Hogan in 1992. It was introduced at Pennsylvania-New Jersey-Maryland (PJM) ISO in the USA. The LMP at a particular node in the system (i.e. bus) signifies the cost of supplying the next increment of load at that node (bus). The price includes the cost of supplying energy marginal cost, cost of losses due to increment in load and cost of losses due to transmission congestion due to the increment. The advantage of LMP mechanism to achieve equilibrium in energy markets is its inherent efficiency in handling network capacity considering the transmission constraints. Also, the price varies from one bus to another bus in the presence of congestion and losses in the network.

Independent System Operator (ISO) determines the locational marginal pricing and power dispatch schedule through a centralised process of market clearing. In the process, network limits are also considered while scheduling generators and loads and bilateral transactions. Since network constraints are considered in the market clearing process, it is not possible to determine the market equilibrium simply by the intersection of a cumulative supply curve and a cumulative demand curve. Hence the power dispatch schedule and LMPs are calculated by modelling the market formulation as an optimisation problem considering network and power flow-related constraints. LMP functions as an indicator of generation deficient regions, load centres and transmission facilities. High value of LMP indicates generation deficient node.

D. Jay and K. Shanti Swarup, *Market Operation for Reactive Power Ancillary Service*, Springer Tracts in Electrical and Electronics Engineering,
https://doi.org/10.1007/978-981-99-6952-4_2

In a single settlement market for LMP scheduling is done for day-ahead. In two settlement market both day-ahead and real-time scheduling are done. Real-time market is within the hour market and done in different time blocks. A 5-min time block is envisaged for real-time markets in PJM ISO. For scheduling in real-time, state estimation solution gives the actual injection by each generator and actual withdrawal by each load at the current point of time.

Calculation of LMP in a market environment involves solving an optimisation problem. LMP is the by-product of the solution to this optimisation problem. In this section, we shall discuss various models existing in literature to calculate LMP.

Locational marginal prices can be arrived at by modelling the market model using two formulation—DC optimal power flow (DCOPF) and AC optimal power flow (AC-OPF). AC-OPF model considers the AC formulation to represent power flow and thus represent the network accurately. However for most of the energy markets especially those not on real-time need not consider such complexities associated with AC power flow. Hence DCOPF models with or without loss components can be considered. In case of a lossless model, the network loss is ignored at the time of LMP calculation. The advantages of DCOPF over AC-OPF for energy markets are its simplicity, speed of convergence and availability of LMP components.

The formulation of lossless DCOPF is given below.

$$\min \sum_{i=1}^{\mathrm{ng}} C_i(P_{gi}) \tag{2.1}$$

subject to

$$\sum_{i=1}^{n} P_{gi} = \sum_{i=1}^{n} P_{di} \tag{2.2}$$

$$-S^{\max} \leq \mathrm{SF} * \left[P_{gi} - P_{di}\right] \leq S^{\max} \tag{2.3}$$

$$P_{gi}^{\min} \leq P_{gi} \leq P_{gi}^{\max} \tag{2.4}$$

where ng is the number of generator offers received
n is the number of buses in the network
M is the number of transmission lines in the network
$C_i(P_{gi})$ is the cost curve of ith generator
P_{gi} is active power generation at ith node
P_{di} is active power demand at ith node
SF is the Sensitivity matrix related to active power injections and line flows
$S^{\max}$ is the maximum active power flow permissible in the transmission line
$P_{gi}^{\min}$ is the lower bound to active power generation
$P_{gi}^{\max}$ is the upper bound to active power generation.

Equation (2.1) is the objective function subject to power flow balance equation given by Eq. (2.2) for lossless model. Equation (2.3) represents the transmission line constraints, and Eq. (2.4) represents the active power generation constraints at each node.

The Lagrangian of the above lossless DCOPF optimisation formulation is given by

$$L_{\mathrm{DCOPF}} = \sum_{i=1}^{\mathrm{ng}} C_i(P_{gi}) - \lambda_{\mathrm{dc}}\left[\sum_{i=1}^{n} P_{gi} - \sum_{i=1}^{n} P_{di}\right] - \mu_{\mathrm{dc1}}\left[\mathrm{SF} * \left[P_{gi} - P_{di}\right] - S^{\max}\right]$$
$$- \mu_{\mathrm{dc2}}\left[P_{gi}^{\min} - P_{gi}\right] - \mu_{\mathrm{dc2}}\left[P_{gi} - P_{gi}^{\max}\right] \tag{2.5}$$

LMP at ith bus is obtained by finding the partial derivative of the Lagrangian with respect to the demand at ith bus which is evaluated as follows:

$$\mathrm{LMP} = \frac{\partial L}{\partial P_{di}} = \lambda_{\mathrm{dc}} + \sum_{k=1}^{M} (\mu_{\mathrm{dc1}k} * \mathrm{SF}_{k-i}) \tag{2.6}$$

Thus the LMP obtained as in Eq. (2.6) from the lossless DCOPF consists of two components: energy component and congestion component. Energy component is related to the power balance equation. The second component corresponds to the inequality constraint corresponding to transmission line flow. In the absence of congestion ($\mu_{\mathrm{dc1}k} = 0$) LMP at all nodes will be the same which corresponds to the energy component considering only the supply–demand balance of classical market equilibrium model.

2.1.1 LMP Calculation in IEEE 9-Bus System

Let us illustrate the calculation of LMP in IEEE 9-bus system using the lossless DCOPF. The IEEE 9-bus is shown in Fig. 2.1.

The system has three generators, and the cost function (quadratic) is as given below

$$C_1(P_{g1}) = 0.11 \cdot P_{g1}^2 + 5 \cdot P_{g1} + 150$$

$$C_2(P_{g2}) = 0.085 \cdot P_{g2}^2 + 1.2 \cdot P_{g2} + 600$$

$$C_3(P_{g3}) = 0.1225 \cdot P_{g3}^2 + 1 \cdot P_{g3} + 335$$

The problem formulation for lossless DCOPF without congestion is written as

$$\min\left(0.11 \cdot P_{g1}^2 + 5 \cdot P_{g1} + 150\right) + \left(0.085 \cdot P_{g2}^2 + 1.2 \cdot P_{g2} + 600\right)$$
$$+ \left(0.1225 \cdot P_{g3}^2 + 1 \cdot P_{g3} + 335\right) \tag{2.7}$$

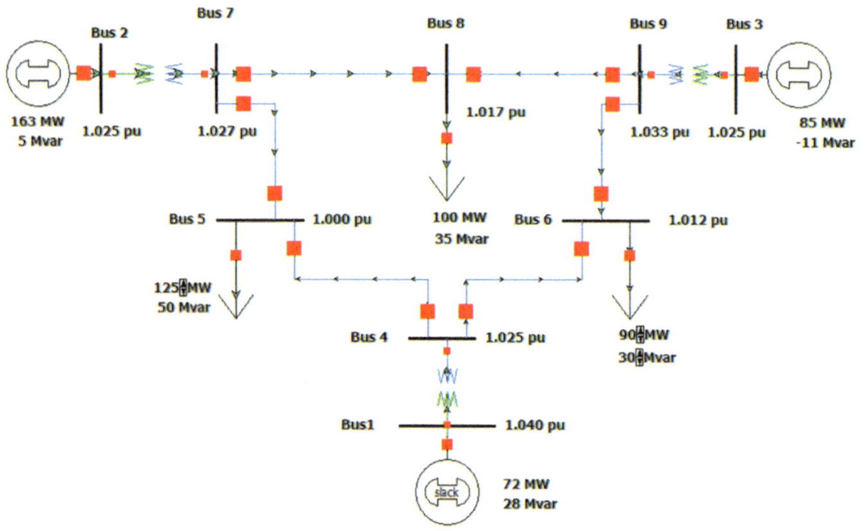

Fig. 2.1 IEEE 9-bus test system

subject to

$$\sum_{i=1}^{3} P_{gi} = \sum_{i=4}^{9} P_{di} \tag{2.8}$$

$$10 \leq P_{g1} \leq 250 \tag{2.9}$$

$$10 \leq P_{g2} \leq 300 \tag{2.10}$$

$$10 \leq P_{g3} \leq 260 \tag{2.11}$$

On solving the lossless DCOPF with no congestion in GAMS platform, the LMP obtained in all three generation nodes is 1.425 \$/MWh. This is because the value of active power produced depends on the Lagrangian variable λ_{dc} only as congestion factor related to transmission lines is not considered.

When more constraints are added to the DCOPF formulation, the value of active power production varies with unit increment in load at any bus. This will be reflected in the LMP values at each node. Each node will have different LMP values depending on the location and severity of network operating conditions.

2.2 Ancillary Services

In traditional vertically integrated utility structure of power systems, generation, transmission and distribution were treated as a single entity owned by a central utility. Thus services like frequency regulation, voltage support, spinning reserves, etc. were bundled along with the essential services and not treated separately. Traditionally, these services were provided by generators through certain obligatory/mandatory measures. Integration of renewable energy sources and the development of smart grid technologies have prompted a shift in the daily grid operation and control. This led to unbundling of certain services from generation. Also, with deregulation, unbundling of service was envisaged which has resulted in identification of two major services in power systems operation: energy supply and ancillary service.

Ancillary services are those unbundled service associated with generation, transmission and distribution which are necessary to support capacity and transmission of electricity while maintaining reliable operation of the grid. These services are required to oblige the network operation constraints within the control area and facilitate continuous flow of electricity to maintain supply–demand balance. In addition to the above primary concern, services that are essential to maintain grid stability and security are also termed as ancillary services that are used to refer to a variety of operations beyond generation and transmission that are required to maintain grid stability and security. These services generally include frequency control, spinning reserves and operating reserves.

Following services have been identified as ancillary services under deregulation:

1. Frequency control ancillary service.
2. Network support and control ancillary service.
3. System restart ancillary service.

2.2.1 Frequency Control Ancillary Service

Frequency control ancillary service (FCAS) is required to ensure the availability of flexible capacity (reserve) to maintain secure operation of the grid during unexpected loss or increase of load, generation or transmission. The objective of frequency control ancillary service is to maintain the grid frequency at any point of time equal to permissible values (e.g. fifty cycles per second).

Frequency control ancillary services can be classified into two depending on the timeline of operation:

1. Regulation frequency control.
2. Contingency frequency control.

Correction of minor generation/demand balance is achieved through regulation frequency control. In contingency frequency control, major generation/demand balance caused by a major contingency event like loss of generation or transmission element is handled. Generally, regulation frequency control is performed in a centralised manner, whereas contingency frequency control is a decentralised control activity, and control operations are triggered locally by the frequency deviation due to contingency.

The regulation frequency control services are provided by generators through Automatic Generation Control (AGC). The AGC system is operated by a centralised grid operator which will monitor the system frequency and send appropriate reference signals for adjusting generation. Contingency services are achieved through local detection of frequency deviation, and balance is achieved through

1. Response of generator governor to frequency deviation.
2. Load shedding.
3. Quick start generation scheduling.
4. Quick unloading of generation unit.

2.2.2 Network Support and Control Ancillary Service

The primary objective of network support and control ancillary service (NSCAS) is to control the system-wide bus voltages within the permissible limit. It is also required to control power flow on transmission network within the limits as per standards. NSCAS will help in maintaining transient and oscillatory stability within the power system after any major event. Thus network support and control ancillary service can be classified into three categories:

1. Voltage control ancillary service.
2. Network loading control ancillary service.
3. Transient and oscillatory stability ancillary service.

Voltage control ancillary service is required to maintain the voltage on the electrical network within permissible limits. Generators either absorb or generate reactive power at their point of connection. This can be achieved through synchronous condensors, dynamic reactive power sources and static reactive power sources.

In network loading ancillary services control of power flow on transmission network is maintained within the limits. This can be controlled through active power flow control elements like phase shifting transformers.

Short-circuit faults may cause sudden spike in power flows which may cause damage to power system equipment through network. This is handled by transient and oscillatory stability ancillary services that control and regulate the network voltage, increase the inertia of the system or rapidly increase/reduce load connected to the power system. Power system stabilisers shall be utilised for this service.

2.2.3 *System Restart Ancillary Service*

It is required to have a proper restoration plan in case of complete or partial blackout of the power system. This is achieved through system restart ancillary services (SRAS). General restart source is considered under system restart ancillary service in which a generator can start and supply to the transmission grid without any external source of supply. In some cases, generators can get disconnected from the main grid to serve internal loads alone upon failure alert. Such generators are restored to supply the main grid only when the system has restored.

2.3 Ancillary Service Markets

Provision of ancillary services is managed directly by Independent System Operator and supplied by grid users according to their licence. Market operations are envisaged for the procurement of ancillary services. There are two ways in which prices can be determined for such ancillary service markets. One method is to determine prices on cost basis. The other method is to arrive at price signals through market mechanisms. Ancillary services shall be procured by ISO through long-term contracts or dedicated markets. Ancillary services are public good and not entitled for individual users. Hence the control of ancillary services is to be under the control of ISO to ensure safe and secure operation of the grid.

It has been identified that financial settlement markets are efficient mode of acquisition for ancillary service markets. For this clear identification of products to be procured through market operations of ancillary service is essential. For example within the operating reserves ancillary service market regulation, spinning reserve and supplemental reserve markets shall be identified separately so that the pricing mechanisms shall reflect the actual requirement in the network.

Since ancillary service markets trade ancillary services that are identified as public goods, the market mechanisms are to be defined in such a way that the price signals are technically and economically efficient. The markets shall incorporate market incentives to participants for providing reliable services. Thus markets can reduce the burden on operators to maintain reliability through out-of-merit actions.

Real-time control of the transmission system has considerable impact on competitive energy markets also. Independent System Operator achieves real-time control by balancing supply and load, congestion management and ancillary services. Hence coordination of energy market with ancillary service markets may be required. One mode of operation is to have sequential clearance of energy and ancillary service markets. Simultaneous clearance of both markets shall also be envisaged. However, simultaneous co-optimisation of energy and ancillary service markets shall function as efficient market coordination mechanism for real-time control of transmission system.

In ancillary service market, ISO/TSO is the operator and purchaser of products under the ancillary service markets. Sellers are generators, reactive power sources, demand response-based prosumers and aggregators. Ancillary service markets may be designed for long-term contracts or spot markets.

In central dispatch process of ancillary service markets, scheduling and dispatch of generation and demand are determined by the ISO. In self-dispatch portfolio-based model, scheduling agents determine the aggregated generation and demand schedules. In self-dispatch unit-based model each generation and demand utility follows their own schedules. In these dispatch processes, ancillary services shall be procured through the following methods:

1. Compulsory or mandatory contracts.
2. Bilateral contracts.
3. Tendering.
4. Spot markets.

In compulsory provision, national regulations and network codes are issued to certain class of generators and are engaged in mandatory contracts to provide ancillary services. The prices are determined through predetermined contracts. Ancillary services are also procured through bilateral contracts in which ISO negotiates with each service provider to purchase specific service quantum. The overall cost is minimised in bilateral contracts, and these are usually placed on long term. However the disadvantage is that the power system dynamics are not included in the bilateral contracts and quantity traded, as these are mostly long-term contracts. Tendering and spot markets induce competition in ancillary service markets.

Pricing mechanism for ancillary services can be based on regulated price, pay-as-bid price or common clearing price. Regulated prices are set by ISO and may not reflect the actual provision cost. In pay-as-bid price scheme, the service provider is paid their offer. In common clearing price, service providers that are successful are paid according to the availability price, utilisation payment and opportunity cost. Availability price refers to the cost of the seller to make a specific amount of ancillary service available. Actual cost associated with the delivery of the product is called utilisation payment. Opportunity cost reflects the possible profit loss in the case where the provider could have sold other products instead of the respective ancillary service provider.

2.4 Challenges in Ancillary Service Markets

Ancillary service being public good is to operated by an independent party like system operator. The quality of the service provided and the price willing to pay for this quality are certain criteria to be considered by the Independent System Operator. To determine proper pricing signal the quality of service should be translated to quantity, quality and location of system service.

However, implementation of ancillary service market faces several challenges. This is because it is difficult to determine the expectation of user. It is also difficult to define the quality of service accurately and the need of the service for power system security and power quality. Thus defining the optimal amount of system service is hard especially when services are shared by areas in interconnected power systems.

The fundamental issues in defining and procuring ancillary services are discussed below:

1. **Defining the responsible entity**
   Ancillary services can be procured by the system operator, or the responsibility can be taken up by users of the system. In PJM market, frequency control services are provided by load-serving entities (LSEs) in pro-rata of their load share. The advantage is that better prices are received due to competition between ancillary service buyers and providers. However, the LSEs do not benefit or use the ancillary service, and hence true value of the service in maintaining the power system dynamics is not evaluated in the market mechanism. Also, the quantity, quality and location of system services are determined by the load-serving entity which is not capable of reducing load.
   When system operator is designated as the responsible entity, the problems like technically efficient price signals and technically optimal operation of the systems can be resolved. However, cost allocation methods are not economically efficient as the priority lies in the technical efficiency alone. Hence suitable regulatory measures are essential to guide the system operator to have economically efficient pricing mechanisms for ancillary service markets.
2. **Lack of investment in ancillary service**
   The revenue expected from ancillary service markets is uncertain. This is because of the nature of requirement of those ancillary services in the network. Hence, providers of ancillary service may not be motivated to invest in improving their capacities and innovate new technologies for ancillary services. This will also result in decreased participation of renewable energy resources, which are potential sources for providing ancillary services like voltage support. Suitable incentive mechanisms are to be envisaged along with ancillary service pricing schemes for rectifying such issues in ancillary service markets.
3. **Defining procurement methodology**
   Several methods for procuring ancillary services were discussed in earlier sections. However, determining appropriate procurement method is influenced by the risk aversion factor of market participants, market concentration, mode of transmission right trading, cost recovery methods, etc. From the market designer perspective, procurement methods that encourage new participants in the market are considered to be suitable. Hence mix of methods is preferred to achieve technically and economically efficient procurement of ancillary services.
4. **Defining bidding structure**
   Since cost associated with providing ancillary service cannot be easily segregated from the energy providing service, there is no specific cost function for service providers when it comes to ancillary services. Accordingly, defining suitable

bidding structure is thus difficult. The price signals are to be issued based on the availability of the service provider, utilisation of the service and also the opportunity cost associated with the service provider.

2.5 Summary and Takeaways

This chapter discussed in brief the two types of electricity markets in deregulated power system, i.e. energy markets and ancillary service markets. The basis of locational marginal pricing mechanism was illustrated through a IEEE 9-bus system in the context of energy markets. The different types of ancillary services were discussed. The challenges in ancillary service markets were described in detail. Identification of challenges in ancillary service markets is essential to understand the complexities in design and implementation of such markets in a real-time network.

Chapter 3
Reactive Power Ancillary Service Markets

Abstract This chapter introduces the concepts of reactive power ancillary service and the basics of reactive power ancillary service markets.

Keywords Reactive power ancillary service · Expected payment function · Obligatory Reactive Power Service · TSO-DSO coordination

3.1 Reactive Power Ancillary Service

In traditional vertically integrated utility (VIU) structure of power systems, generation, transmission and distribution were considered as a single bundled service owned by a central entity. With deregulation or unbundling in power systems, these services are separated, and sale of these services is achieved at different rates through market operations. Thus, reactive power management and voltage support that were earlier bundled with other services in VIU structure are now identified as one of the most important ancillary services (Commission et al. 1996). The advantages of reactive power support ancillary service market operations are (Deksnys and Staniulis 2007):

1. Motivating stakeholders to install reactive power sources.
2. Compensating individual generators for reactive power costs associated in providing additional voltage support in the system.
3. Optimal scheduling of reactive power by operators that ensure safe and secure grid operation.

Reactive power is an integral part of alternating current (AC) power systems due to the reactive (inductive or capacitive) nature of electrical loads, transmission and distribution networks. Circulation of reactive power in the network results in excess current flow thereby increasing losses, and hence reactive power demand is to be satisfied locally. If sufficient reactive power is not supplied in the network, system-wide bus voltages may collapse. Overhead transmission lines absorb reactive power when loaded above their Surge Impedance Loading (SIL), while they generate reactive power during light load conditions. Underground cables are always loaded below their SIL and hence act as reactive power sources. Thus even if reactive power

D. Jay and K. Shanti Swarup, *Market Operation for Reactive Power Ancillary Service*, Springer Tracts in Electrical and Electronics Engineering,
https://doi.org/10.1007/978-981-99-6952-4_3

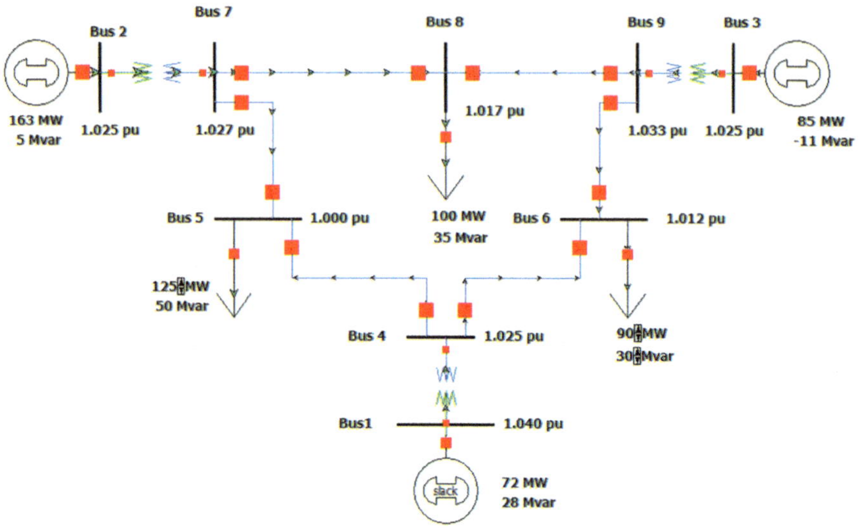

Fig. 3.1 IEEE 9-bus system base case

demand from loads is satisfied through local reactive power sources, transmission network consumes reactive power during peak real power demand in the system. Hence reactive power is required for the shipment of active power in the network to maintain system-wide bus voltages within the permissible limits.

The importance of reactive power management is illustrated in a IEEE 9-bus system. The example shows the impact of location of source, active power loading and system operating condition on reactive power management. The base case of IEEE 9-bus system is shown in Fig. 3.1.

On the same loading conditions, the operating conditions when generation at Bus 3 exercises MVAr withholding are shown in Fig. 3.2.

It can be observed that on MVAr withholding at Bus 3, the system bus voltages increase. The total voltage deviation (Ahmadi and Foroud 2013) in this case considering base case voltage as reference is 0.03198. This indicates the relevance of effective reactive power management in the network.

To illustrate the impact of location of reactive power source in the network, consider the base case loading condition with MVAr withholding at Bus 2. This is shown in Fig. 3.3.

The total voltage deviation (Ahmadi and Foroud 2013) considering base case voltage as reference in this case is 0.01411. On comparing with the case of MVAr withholding at Bus 3, it is observed that the MVAr source at Bus 3 is more important in maintaining a better voltage profile. Thus location of source has an increased impact on management of reactive power in the network.

To demonstrate the inherent coupling between active and reactive power market operations, consider the case when load at Bus 5 increases by 200 MW. The operating scenario in this active power load increment case with all reactive power sources

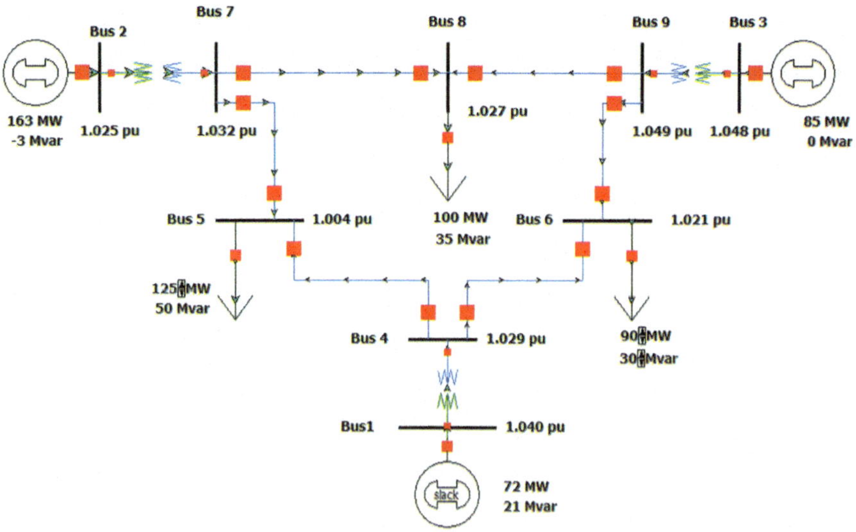

Fig. 3.2 IEEE 9-bus system with base case loading and MVAr withhold at Bus 3

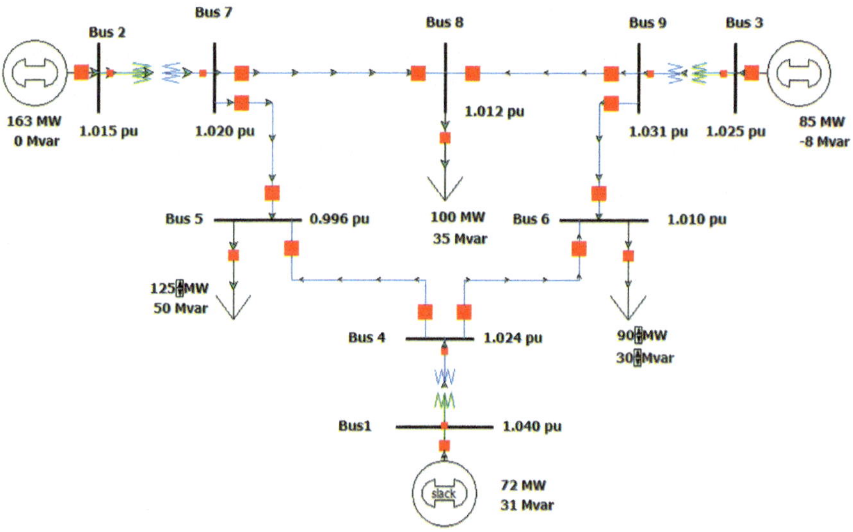

Fig. 3.3 IEEE 9-bus system base case—with MVAr withhold at Bus 2

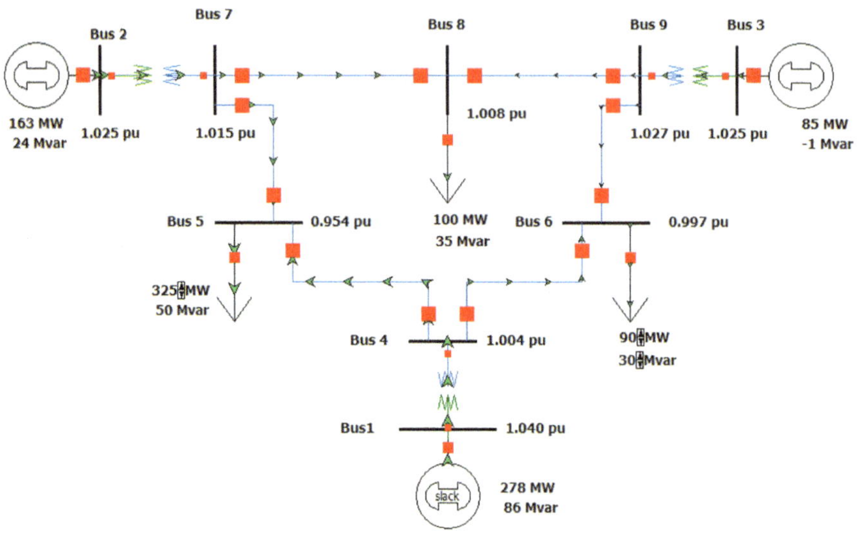

Fig. 3.4 IEEE 9-bus system—200 MW load increment at Bus 2

providing voltage support is shown in Fig. 3.4. The net MVAr generation IEEE 9-bus system (base case) is 22 MVAr. With 200 MW load increment in the system, MVAr generation required for maintaining voltage profile within the permissible limits is 109 MVAr. This indicates that the inherent coupling between active and reactive power market caused by network properties is to be considered in addition to the capability curve considerations of synchronous generators for efficient reactive power management in the system.

3.2 Basic Concept of Reactive Power Markets

A generic flowchart depicting the process flow in reactive power market mechanism based on nodal pricing is given in Fig. 3.5. Price signals are issued to market participants in reactive power ancillary service market by Independent System Operator (ISO). ISO decides this by identifying local market areas so as to handle the localised nature of reactive power requirement in the system. Reactive power market is prone to exercise of market power, and these indices are to be considered while determining the price signals. The need for reactive power in the system is determined by the voltage security constraints and uncertainty in demand and renewable sources. Hence these factors are to be incorporated while solving AC optimal power flow (AC-OPF). The basic formulation of AC-OPF is as below
Minimise

$$\sum_{i \in \mathbb{G}} C(Q_{gi}) \tag{3.1}$$

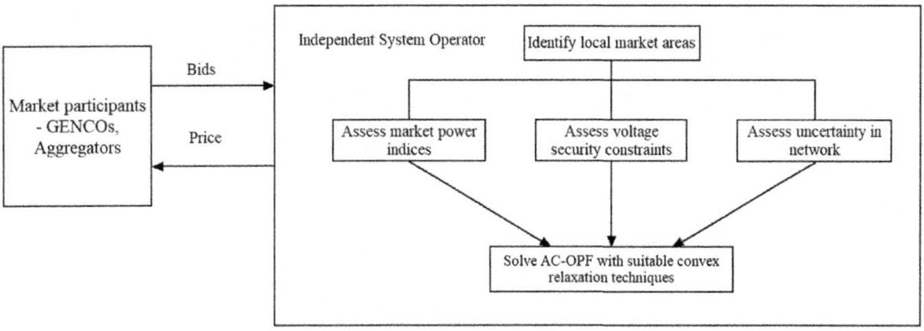

Fig. 3.5 Reactive power market mechanism

Subject to:

Equality constraints

$$P_{gi} - P_{di} = \sum_{j \in \mathbb{N}} |V_i||V_j|Y_{ij} \cos\left(\theta_{ij} + \delta_j - \delta_i\right) \tag{3.2}$$

$$Q_{gi} - Q_{di} = -\sum_{j \in \mathbb{N}} |V_i||V_j|Y_{ij} \sin\left(\theta_{ij} + \delta_j - \delta_i\right) \tag{3.3}$$

Inequality constraints

$$P_{gi}^{\min} \leq P_{gi} \leq P_{gi}^{\max} \tag{3.4}$$

$$Q_{gi}^{\min} \leq Q_{gi} \leq Q_{gi}^{\max} \tag{3.5}$$

$$-S_l^{\max} \leq S_l \leq S_l^{\max} \tag{3.6}$$

$$V_i^{\min} \leq V_i \leq V_i^{\max} \tag{3.7}$$

where $C(Q_{gi})$ is the objective function of AC-OPF formulation which in most of the cases considered as the cost of reactive power generation. The constraints in AC-OPF include equality constraint for power flow equations over the transmission network expressed in polar form. Inequality constraint set includes the active and reactive power generation limits as well as the transfer capability of transmission lines. Voltage magnitude limits are also considered in AC-OPF.

Since it is difficult to decouple the cost of active-reactive power generation and as fuel cost is mostly associated with active power generation, separate payment functions have been formulated for reactive power procurement. The payment function that has been widely used in literature is expected payment function (EPF) which is based on the reactive power capability curve of synchronous generators as shown in Fig. 3.6 (Zhong and Bhattacharya 2002).

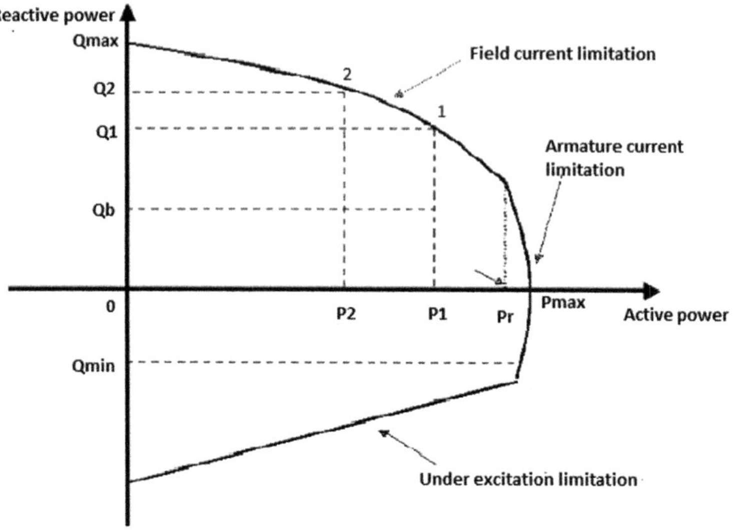

Fig. 3.6 Capability curve of synchronous generator

Fig. 3.7 Offer from reactive power market participants (Zhong and Bhattacharya 2002)

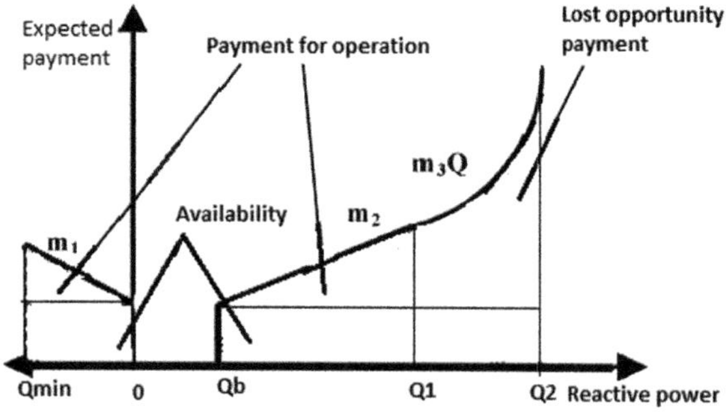

In Fig. 3.6, when the operating point is at (P_1, Q_b), the generator can increase the reactive power generation up to Q_1 without reducing active power production (P_1). However an increase in reactive power generation/absorption increases the real power loss in the windings. This is termed as operation cost. If the system requirements are such that additional reactive power support is to be provided, then active power production is to be reduced from P_1 to P_2. This results in a financial loss to generating companies in active power market which is to be compensated in reactive power market. This is termed as lost opportunity cost. This expected payment function suits for synchronous condensers as well, where in the lost opportunity cost offer from condensers is zero. Thus the reactive power sources are to be paid for these bands of operation. Hence the expected payment function (EPF) is defined as (Zhong and Bhattacharya 2002) with structure of reactive power offer from participants as in Fig. 3.7.

$$\text{EPF} = \text{Availability cost} + \text{Operation cost} + \text{Lost opportunity cost}$$

$$\text{EPF}_i = a_{0i} + \int_{Q\text{min}}^{0} m_{1i}\mathrm{d}Q_i + \int_{Q_b}^{Q_1} m_{2i}\mathrm{d}Q_i + \int_{Q_1}^{Q_2} m_{3i}\, Q_i\mathrm{d}Q_i \tag{3.8}$$

Thus when market participants submit their availability cost offer (a_{0i}), operation cost in absorption mode (m_{1i}), operation cost in generation mode (m_{2i}) and lost opportunity cost (m_{3i}), the expected payment function can be written as in Eq. (3.8).

3.3 Existing Reactive Power Ancillary Service Markets

Recognising the relevance of reactive power ancillary service market in achieving safe and secure operation of the grid, mechanisms suitable for short-term and long-term contracts have been designed and implemented world wide. The global experience in the implementation of reactive power pricing is detailed in Stucchi and Pollitt (2018). Australian Energy Market Operator (AEMO) incorporates tender-based market structure for reactive power procurement through voltage control ancillary service market. The contracts may be short term, long term or a combination of both. The contracted reactive power is evaluated through two modes of operation—capacitive and inductive. Payment involves availability and compensation when the generator units are constrained. In Great Britain, three mechanisms, namely Obligatory Reactive Power Service (ORPS), Enhanced Reactive Power Service (ERPS) and Transmission Constraint Management (TCM) are used to acquire reactive power ancillary services. ORPS mechanism is a mandatory service. ERPS is acquired through tenders that are held every six months. TCM is based on bilateral contracts depending on system operating conditions. In California ISO (CAISO) reactive power pricing is based on opportunity cost. A fixed rate payment is implemented for reactive power ancillary service in New York ISO (NYISO), while American Electric Power (AEP) payment method is utilised in PJM. A summary of reactive power procurement methods adopted across the globe is given in Table 3.1.

In India, according to Indian Electricity Grid Code Section 6.6, regional entities are bestowed with the responsibility of managing reactive power locally. Generation companies are not involved in the pricing methodology. The pricing formulation is as follows:

1. When voltage at metering point is below 97%, the regional entity pays for VAr drawl at Inter State Transmission System (ISTS).
2. When voltage at metering point is below 97%, the regional entity gets paid for VAr return at (ISTS).
3. When voltage at metering point is above 103%, the regional entity pays for VAr return at ISTS.

Table 3.1 Survey of reactive power procurement methods

System operator	Mechanism	Price components
CAISO (USA)	Mandatory service	Opportunity cost
PJM, NYISO (USA)	Mandatory service	1. Capability payment 2. Opportunity cost
AEMO	1. Generation mode through tender 2. Synchronous condensor mode through tender	1. Generation mode a. Availability cost b. Opportunity cost c. Testing charges, cost for voltage support 2. Synchronous condensor mode a. Enabling cost b. Testing charges, cost for voltage support
National grid (Great Britain)	1. Obligatory Reactive Power Service 2. Enhanced Reactive Power Service—through tender	1. Utilisation payment in Obligatory Reactive Power Service 2. Enhanced Reactive Power Service a. Availability cost b. Utilisation

4. When voltage at metering point is above 103%, the regional entity is paid for VAr drawl at ISTS.

Charge of 10 paise/KVArh with effect from 03.05.2010 was decided with an escalation rate of 0.5 paise/KVArh per year. A survey on the southern region grid based on the operational committee coordination minutes was carried out, and it was observed that in spite of the predetermined pricing for reactive power, reactive power injection has been observed at nodes resulting in high-voltage profile of the grid. The issue of localised high voltage/low voltages still persisted in the southern region grid. In October–December (2019–2020) quarter total number of lines hand tripped due to over voltage was 158. It was noticed that Vallore–Alamathy Circuit 1 was hand tripped 12 times with total outage hours 361.6 h in that quarter due to over voltage issues. In July–September (2019–2020) quarter 141 lines were hand tripped due to over voltage issues.

Reactive power injection was observed from lines between Vallore TPS–Alamathy and Vallore TPS–Kalivanthapattu. The reactive power injection/absorption at Vallore TPS was not well coordinated, and with the existing pricing mechanism, Tamil Nadu was payable to the pool for reactive power injection during high-voltage profile.

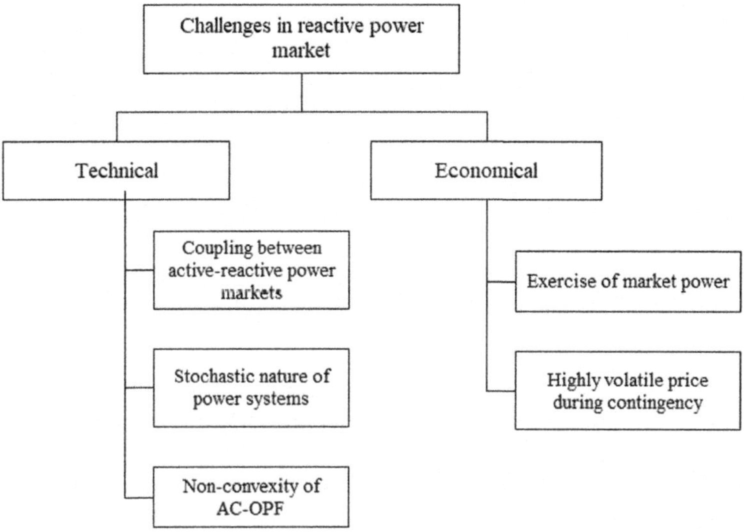

Fig. 3.8 Challenges in implementing reactive power market

Generators' contribution/participation played a very significant role in voltage management. It was noted that few units did not keep reference set value for voltage at nominal levels and were being changed manually to control/restrict reactive power interchange. The coordination committee has decided measures like check of reactive power capability curve at each generator plant, optimisation of generator and interconnecting transformer tap positions, mandatory operation at synchronous condenser mode, etc. However, these measures may fail when there is an increased participation of private generation companies in grid operation. Even though solar generators and wind generators have very controllable reactive injection/absorption possibility they are not being utilised because of Power Purchase Agreement commitments and penalties. Hence, suitable pricing mechanism is required for reactive power ancillary service that ensures incentives to compensate for such commitments and penalties to ensure safe and secure operation of grid.

3.4 Challenges in Reactive Power Market

Nodal prices that are correlated to the Lagrangian dual variables of AC optimal power flow (AC-OPF) are considered to be an effective pricing mechanism as the system operating conditions are well incorporated. Also the localised nature of reactive power can be captured well when nodal pricing mechanism is considered. However implementation of nodal prices for real-time reactive power market faces several challenges that are summarised in Fig. 3.8.

Uncertainty in power systems

Reactive power demand from loads is stochastic in nature especially when price-based demand response schemes are involved. This affects the implementation of hourly prices determined through day-ahead-based market mechanism. Also an increased participation of renewable energy sources in reactive power market results in uncertainty of reactive power generation as well. It may also be noted that system contingencies and increased active power generation affect reactive power pricing, and hence nodal pricing-based markets (especially in day-ahead settlement) face challenges due to uncertainty in system topology and power flows.

Non-convexity of AC-OPF

Nodal prices are based on AC-OPF. However AC-OPF is a non-convex, nonlinear problem. Such problems are generally NP-hard, and as a result globally optimal solutions are not guaranteed. Also the distance to global optimum solution cannot be specified. Hence convexification techniques will be required to achieve desired tractable solutions during all contingencies. However this may result in increased computational time and efforts which will affect market operations.

Inherent coupling between active and reactive power

In the case of synchronous generators, the coupling between active and reactive power is depicted through the capability curve as in Fig. 3.6. When reactive power generation has to be increased for providing required support in the system and shipment of active power, generators may be required to decrease active power production. This coupling has though been treated in payment function as lost opportunity cost, it is difficult to quantify the actual cost associated with this phenomenon. Also, when active power demand from loads increases with reactive power demand being constant, in order to maintain voltages at bus within the permissible limits, additional reactive power generation is required. This shows that the active and reactive power markets influence each other, and hence the market mechanism design must consider this inherent coupling nature while defining a pricing structure for reactive power.

Price volatility

Reactive power prices are evaluated based on the requirement of reactive power in the system. Requirement of reactive power in the system depends on the network topology, operating conditions, etc., and hence the prices are highly correlated with the system contingencies. Thus if nodal price-based reactive power market structure is considered, prices may be highly volatile. Localised nature of reactive power also may result in prices at some locations to be too high depending on system operating conditions. However, a uniform pricing scheme may not be a technically efficient pricing signal as it may fail to encourage reactive power sources at critical locations to provide reactive power support in the system. Hence, volatility in nodal prices is to be curtailed to avoid unnecessary burden on customers.

Market power

Reactive power is desired to be supplied locally to avoid transmission through long transmission lines and thereby reduce loss in the system. This fundamental feature of reactive power support provides opportunities for market participants to exercise market power. Market power is the ability of market participants to earn profit by altering price beyond competitive level (Twomey et al. 2005). This may arise mainly due to the limited number of potential suppliers in the system. Also, with inherent coupling between active and reactive power, market participants may tend to withhold reactive power generation, thereby earning increased prices in active power market and vice-versa (Ganger et al. 2013). When there is congestion in the system, sources near load centres may alter prices much higher than their competitive level. Thus a market mechanism for reactive power must require strong measures to mitigate market power.

3.5 Ideal Reactive Power Market

Reactive power support ancillary service market is necessary to achieve required voltage support in the system. The grid operator is responsible for reliability of the system, and these operations need reactive power in real-time to deal with contingencies. Also, investment is required to meet the demand which is to be scheduled on a day-ahead basis. Non-market strategies are effective through mandatory services wherein there is transparency regarding the reactive power capability of each source. Such non-market strategies are enforced to be effective through periodic inspection and testing of capability curves of sources. This is effective in cases where most of the reactive power sources are state-owned.

However, such mechanisms will discourage investment of private firms in reactive power market. Also, non-market strategies may not ensure that the private firms are well compensated to provide reactive power support under contingencies. This is because the price signals in non-market strategies are not competitive and private firms are not encouraged to invest in reactive power support schemes. Renewable energy sources are identified as potential sources for reactive power especially during contingencies and provide real-time reactive power support. However, the solar and wind power plants are not actively involved in providing reactive power support due to commitments and penalties associated with the Power Purchase Agreement. On devising suitable market strategies, an efficient price signal shall be arrived at, which will encourage private firms to invest in reactive power support and also render their service in providing real-time reactive power support.

Market structures for reactive power support will enable to solve the aforementioned issues in achieving reactive power support in a deregulated system with increased penetration of renewable energy sources. This is accomplished by issuing proper price signals to the market participants that submit their bids to provide

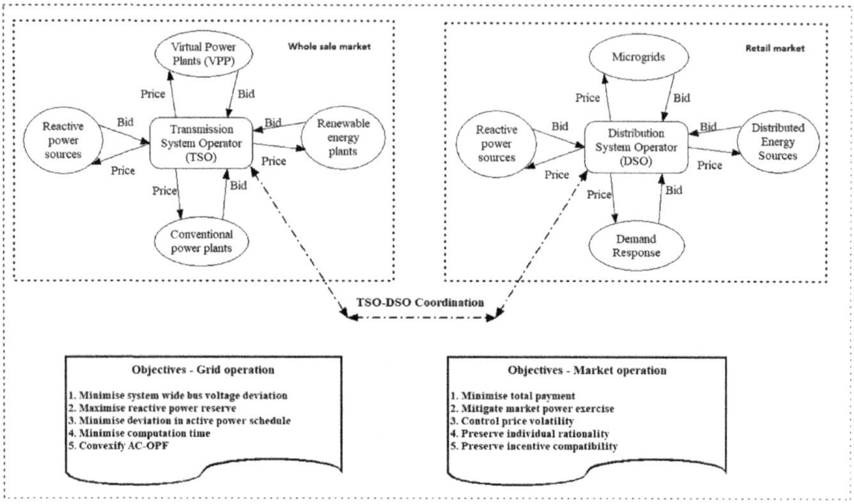

Fig. 3.9 Framework for reactive power ancillary service market

reactive power support. A framework for reactive power support ancillary service market is as depicted in Fig. 3.9 (Jay and Swarup 2021).

The key features of an ideal reactive power market mechanism are

1. Mitigate market power.
2. Control price volatility.
3. Preserve individual rationality.
4. Preserve incentive compatibility.
5. Optimal reactive power scheduling through optimal payment.

The framework for reactive power ancillary service market provided in the paper represents an ideal reactive power market structure suitable for smart grid under deregulation. Existing technologies that have been implemented for other market operations and control strategies are extended to reactive power ancillary market, and the coordination techniques are depicted in the framework proposed in this paper. The authors are of the opinion that a research initiative in this direction will facilitate the implementation of a real-time reactive power ancillary service market utilising the existing TSO-DSO coordination techniques and active power market technology. The objectives that the system operator achieves through reactive power market that are to be considered in formulating a reactive power market are

1. Minimise total voltage deviation.
2. Maximise reactive power reserve.
3. Minimise deviation in active power scheduling.

In reactive power ancillary service market, two levels of market exist—wholesale and retail markets. Transmission system operator (TSO) received bids from market participants and issue price signals to them in the wholesale market to achieve required reactive power support in the system. Conventional power plants, large-scale renewable energy plants, independent reactive power sources and virtual power plants/aggregators are the major market participants in reactive power market at a wholesale level, i.e. transmission levels.

Distribution system operator (DSO) issues price signals to market participants in the retail market so that required reactive power scheduling is achieved. Micro-grids, prosumers, customers with price-based demand response strategies, distributed energy resources and reactive power sources participate in reactive power ancillary service market. This will help DSO to maintain bus voltages within the permissible limits.

The techniques that are to be implemented in achieving an ideal reactive power market are

1. Network partitioning for identifying local market areas or relevance of market participants.
2. Convexification of AC optimal power flow (AC-OPF) so as to improve computational time and effort suitable for real-time pricing mechanisms.
3. Stochastic AC-OPF formulation to consider network uncertainties.
4. Proper TSO-DSO coordination schemes for achieving stable market operations.

With advanced communication technology, increased penetration of distributed energy resources and more demand response schemes, active distribution networks play a significant role in maintaining system-wide bus voltages at the main grid-level under TSO. Hence proper coordination between TSO and DSO is required in reactive power markets as well so that the prices in both wholesale and retail power markets are stable, and optimised scheduling of reactive power at sources is achieved. DISCOs serve as customers as well as providers. Hence distribution system operator (DSO) should be well coordinated with transmission system operator (TSO) in reactive power market. Thus TSO-DSO coordination schemes are important in the implementation of an efficient reactive power market to consider the prosumer aspect of DISCOs pertaining to reactive power.

Implementation of the proposed reactive power market framework will involve proper utilisation and coordination of existing technologies. To identify local market areas, it is required to partition the network. For defining such local market areas that adapt with real-time system operating conditions, the bus admittance matrix can be estimated from PMU measurements. From the bus admittance matrix, relative electric distance measure between buses can be defined using techniques that are computationally efficient. Solution of AC-OPF will require suitable non-convex optimisation algorithm using parallel computing with efficient computational machines for real-time applications. To handle uncertainties, suitable studies on load forecasting, renewable energy forecasting, etc., are also required for formulating stochastic AC-OPF. Efficient communication technologies suitable to achieve TSO-DSO coordination are also to be studied in detail. The topology and protocol for communication will depend on the schemes for coordinating TSO and DSO. The advantage is that the schemes that will suit TSO-DSO voltage coordination in real-time may also suit the reactive power coordination problem.

For efficient implementation of reactive power market, it is required that a dedicated team shall perform a detailed study on the behaviour of reactive power market in each local area. This is important to consider the localised nature of reactive power and at the same time limit the circulation of reactive power in the network. These studies are also important to provide feedback to network expansion planning group, so that investment in reactive power source installation shall be envisaged so as to avoid monopoly of other sources in the local area. A detailed study on the bidding strategies of market participants in reactive power market shall also be carried out. This will help in defining suitable market policies and regulatory measures which will improve the bidding structure and market efficiency.

References

Ahmadi H, Foroud AA (2013) A stochastic framework for reactive power procurement market, based on nodal price model. Int J Electr Power Energy Syst 49:104–113

Commission UFER et al (1996) Promoting wholesale competition through open access non-discriminatory transmission services by public utilities; recovery of stranded costs by public utilities and transmitting utilities. Order 888:24

Deksnys R, Staniulis R (2007) Pricing of reactive power service. Oil Shale 24

Ganger D, Zhao J, Hedayati M, Mandadi A (2013) A review and simulation on real time reactive power spot markets. In: 2013 North American power symposium (NAPS). IEEE, pp 1–5

Jay D, Swarup K (2021) A comprehensive survey on reactive power ancillary service markets. Renew Sustain Energy Rev 144:110967

Stucchi KA, Pollitt M (2018) Reactive power procurement: lessons from around the world. In: Evolving energy realities: adapting to what's next, 36th USAEE/IAEE North American conference, 23–26 Sept 2018. International Association for Energy Economics

Twomey P, Green R, Neuhoff K, Newbery D (2005) A review of the monitoring of market power. Massachusetts Institute of Technology Center for Energy and Environmental Policy Research. http://ideas.repec.org/p/cam/camdae/0504.htmlConsultadaelel29

Zhong J, Bhattacharya K (2002) Toward a competitive market for reactive power. IEEE Trans Power Syst 17(4):1206–1215

Part II
Design of Reactive Power Markets

Chapter 4
Reactive Power and Network Partitioning

Abstract This chapter presents the concepts of partitioning the network into local areas, to capture the localised nature of reactive power requirement in defining market models.

Keywords Localised reactive power · Graph partitioning · Voltage control areas · Voltage-apparent power coupled areas

A discussion on the relevance of network partitioning technique in reactive power markets is presented in this chapter. Network partitioning techniques and algorithms suitable for reactive power markets are detailed in this chapter.

4.1 Relevance of Network Partitioning Techniques

The implementation of reactive power markets must incorporate the localised requirement of reactive power in the network. This is because, circulation of reactive power in the network is to be avoided to reduce losses. Hence reactive power requirements in the network are to be satisfied locally. This basic and important system requirement has imposed several challenges in the implementation of reactive power markets.

One major challenge that arises in reactive power markets due to the localised requirement of reactive power is that reactive power support service providers have the potential of exercising market power. Market power is the ability of market participants to alter the prices away from competitive levels profitably. The exercise of market power may be intentional or unintentional. Quantity withholding and financial withholding are the main strategies in exercising market power. In addition to these, transmission-related strategies like creating congestion, dispatch of generation deviating from marginal cost, etc., are also identified as strategies for exercising market power in electricity markets.

Localised requirement of reactive power may increase the locational marginal prices obtained through AC-OPF in certain nodes especially in the cases of system contingencies. Hence price volatility due to intentional or unintentional contingencies

D. Jay and K. Shanti Swarup, *Market Operation for Reactive Power Ancillary Service*, Springer Tracts in Electrical and Electronics Engineering,
https://doi.org/10.1007/978-981-99-6952-4_4

and congestions in the network is another challenge that impairs the implementation of reactive power markets.

One solution methodology adopted to handle these challenges that arise due to localised nature of reactive power is to define local market areas in the large network. The advantages of defining the local market areas for reactive power markets are that

1. The effect of price alteration due to exercising of market power can be controlled within a smaller area. This will reduce the impact of market power on the entire system and enable an improved economic operation of the market.
2. Price volatility due to intentional/unintentional contingencies and congestion can be curtailed to an extent using local market areas especially when zonal prices are implemented.
3. Optimal allocation of reactive power sources can be achieved so that the reactive power requirement can be satisfied locally to avoid circulation of reactive power in the network.

How to define local market areas for reactive power market? The primary objective of reactive power ancillary service is to maintain the system-wide bus voltages within the permissible limits. Hence while defining local market areas, voltage control aspects are to be considered. By suitable defining control areas based on the electrical distance measure that relates the voltages between buses, the large network shall be partitioned into local market areas. Based on the electrical distance measure or criteria for defining control areas, suitable partitioning technique shall be employed for partitioning the network. Techniques that can be employed for partitioning large networks can be broadly classified into

1. Graph partitioning algorithms (Mehrjerdi et al. 2012).
2. Heuristic algorithms (Bahmanifirouzi et al. 2012).
3. Machine learning (data-driven) clustering techniques (Abou Daher et al. 2013).

However, these techniques are to be studied in detail to consider the computational efforts involved in the implementation of reactive power markets for large networks. This will depend on the system configuration and tools available at the market operator for implementation of localised reactive power markets. Irrespective of the technique used, the basis of partitioning is the definition of electrical distance measure in the network. This definition is particularly important as it serves the basis for partitioning the network into local market areas. Electrical distance between two buses can be quantified through the bus admittance matrix of the system and from the Jacobian matrix of the converged load flow solution. Hence electrical distance measure is to be defined in such a way that the local market areas defined for reactive power market provide required voltage support in the network and improve the grid voltage profile. The definition should also be capable of curtailing the impact of market power exercised by certain participants on the entire grid. Price volatility due to contingencies shall also be mitigated and curtailed within the local area upon suitable definitions of electrical distance measure.

4.2 Voltage Control Areas

The system of equations that govern the power system network with N number of buses is given by

$$P_i = \sum_{j=1}^{N} V_i V_j Y_{ij} \cos(\delta_i - \delta_k + \theta_{ik}) \tag{4.1}$$

$$Q_i = -\sum_{j=1}^{N} V_i V_j Y_{ij} \sin(\delta_i - \delta_k + \theta_{ik}) \tag{4.2}$$

The above nonlinear equations shall be solved using Newton Raphson method, in which the linearised model shall be represented as

$$\begin{bmatrix} \Delta P \\ \Delta Q \end{bmatrix} = \begin{bmatrix} P_\delta & P_V \\ Q_\delta & Q_V \end{bmatrix} \begin{bmatrix} \Delta V \\ \Delta \delta \end{bmatrix} \tag{4.3}$$

The matrix $\begin{bmatrix} P_\delta & P_V \\ Q_\delta & Q_V \end{bmatrix}$ is called the Jacobian matrix where $P_\delta = \frac{\partial P}{\partial \delta}$; $Q_\delta = \frac{\partial Q}{\partial \delta}$; $P_V = \frac{\partial P}{\partial V}$; $Q_V = \frac{\partial Q}{\partial V}$.

The inverse of Q_V is called the sensitivity matrix which is real and non-symmetrical. The elements of the sensitivity matrix represent the propagation of voltage variation following reactive power injection at a bus. The magnitude of voltage coupling between two buses can be quantified by the maximum attenuation of voltage variation between these two buses. The magnitude of voltage coupling between two buses can be quantified by the maximum attenuation of voltage variation between these two buses. The attenuation values are obtained by dividing the elements of each column of the Q_V matrix by the diagonal term. Thus, Zhong et al. (2004)

$$\Delta V_i \cdot \frac{\partial Q}{\partial V_k} = \frac{\partial Q}{\partial V_i} \cdot \Delta V_k \tag{4.4}$$

The attenuation factor is defined as

$$\alpha_{ik} = \frac{\partial V_k}{\partial Q_k} / \frac{\partial V_i}{\partial Q_k} \tag{4.5}$$

Generally, $\alpha_{ik} \neq \alpha_{ki}$. Hence for positive symmetric weight for the network, electric distance is defined as

$$D_{ik} = D_{ki} = -\log(\alpha_{ik}, \alpha_{ki}) \tag{4.6}$$

The steps involved in determining the electrical distance are

1. Determine the sensitivity matrix from Jacobian.
2. Find the attenuation factors as given in Eq. (4.5).
3. Calculate the positive symmetric matrix for electrical distance by Eq. (4.6).

4.2.1 Hierarchical Classification Algorithm

To partition the network into voltage control areas for defining local market areas for reactive power market, hierarchical classification algorithm is widely used. According to this methodology, the electrical distance from ith generator with other generators is classified into ranges.

Let the range limits be $\mathbf{A} < \mathbf{B} < \mathbf{C}$. Then for buses with electrical distance less than **A** falls in classification 1. For buses having electrical distance between **A** and **B** falls in classification 2. Those buses with electrical distance between **B** and **C** are under classification 3.

It has been observed that the smaller the range limits, the better the classification. The overlaps between groups can be minimised for smaller range limits, thereby achieving non-overlapping partitions in the network. This will be suitable when control areas or local market areas are defined.

4.2.2 Spectral Clustering

From Eq. (4.3), assuming change in active power to be zero

$$\Delta Q = (Q_V - Q_\delta P_\delta^{-1} P_V)\Delta V \tag{4.7}$$

This defines the volt/var coupling between buses in the network, which is defined as the weights of graph that represents the network. Hence graph partitioning techniques like spectral clustering can be applied on the network with weights given by the volt/var coupling factor in Eq. (4.7). The step involved in forming local market areas through volt/var coupling factor using spectral clustering technique is as given below (Jiang et al. 2017):

1. Determine the volt/var coupling factor between buses in the network as given by Eq. (4.7).
2. Determine the graph representation of the network with weights given by the volt/var coupling factor.
3. Find the normalised Laplacian matrix of the weighted graph.
4. Calculate the eigenvalues of the normalised Laplacian matrix and find the maximum relative eigengap to determine k, i.e. the number of clusters of the graph.
5. Normalise the first k eigenvectors of the normalised Laplacian matrix.

6. Apply the spectral clustering to the first k eigenvectors of the graph Laplacian matrix to partition the graph into k clusters.

4.3 Relative Electrical Distance

Consider a system with g number of generator buses and $(b–g)$ number of load buses, where b is the total number of buses in the system. Then for the given system,

$$\begin{bmatrix} I_{\mathrm{L}} \\ I_{\mathrm{G}} \end{bmatrix} = Y_{\mathrm{bus}} \begin{bmatrix} V_{\mathrm{L}} \\ V_{\mathrm{G}} \end{bmatrix} \tag{4.8}$$

where $Y_{\mathrm{bus}} = \begin{bmatrix} Y_{\mathrm{LL}} & Y_{\mathrm{LG}} \\ Y_{\mathrm{GL}} & Y_{\mathrm{GG}} \end{bmatrix}$.
$\boldsymbol{I}_{\mathrm{L}}$ and $\boldsymbol{V}_{\mathrm{L}}$ are the vectors that represent current and voltage phasors of load buses, and $\boldsymbol{I}_{\mathrm{G}}$ and $\boldsymbol{V}_{\mathrm{G}}$ are vectors that represent the current and voltage phasors at generator buses. Equation 4.8 is rewritten as

$$\begin{bmatrix} I_{\mathrm{G}} \\ I_{\mathrm{L}} \end{bmatrix} = \begin{bmatrix} Y_{\mathrm{GG}} & Y_{\mathrm{GL}} \\ Y_{\mathrm{LG}} & Y_{\mathrm{LL}} \end{bmatrix} \begin{bmatrix} V_{\mathrm{G}} \\ V_{\mathrm{L}} \end{bmatrix} \tag{4.9}$$

The aforementioned systems of equations have been rearranged in Visakha et al. (2004) as

$$\begin{bmatrix} V_{\mathrm{L}} \\ I_{\mathrm{G}} \end{bmatrix} = \begin{bmatrix} Z_{\mathrm{LL}} & F_{\mathrm{LG}} \\ K_{\mathrm{GL}} & Y_{\mathrm{GM}} \end{bmatrix} \begin{bmatrix} I_{\mathrm{L}} \\ V_{\mathrm{G}} \end{bmatrix} \tag{4.10}$$

where
$\boldsymbol{Z}_{\mathrm{LL}} = \boldsymbol{Y}_{\mathrm{LL}}^{-1}$
$\boldsymbol{F}_{\mathrm{LG}} = -\boldsymbol{Y}_{\mathrm{LL}}^{-1}\boldsymbol{Y}_{\mathrm{LG}}$
$\boldsymbol{K}_{\mathrm{GL}} = \boldsymbol{Y}_{\mathrm{GL}}\boldsymbol{Y}_{\mathrm{LL}}^{-1}$
$\boldsymbol{Y}_{\mathrm{GM}} = \boldsymbol{Y}_{\mathrm{GG}} - \boldsymbol{Y}_{\mathrm{GL}}\boldsymbol{Y}_{\mathrm{LL}}^{-1}\boldsymbol{Y}_{\mathrm{LG}}$.

F_{LG} matrix is complex with columns corresponding to the generator bus numbers and rows corresponding to the load bus numbers. The elements of the matrix give the relation between load-generation bus voltages. The absolute value of F_{LG} matrix provides the ideal generation proportions. This value is termed as relative electrical distance (RED). Visakha et al. (2004) gives the information about the location of load nodes with respect to generator nodes.

4.4 Voltage-Apparent Power Coupling Factor

In this section, the relative electrical distance measure is modified to propose a novel relative electrical distance measure that determines the voltage-apparent power coupling between load buses and load-generation buses. The measure is derived from the bus admittance matrix. It is assumed that the bus admittance matrix is known, either directly from the system topology or estimated using PMU-based state estimation techniques in cases where system topology is not known. The optimal number of PMUs required for minimising error in state estimation depends on the observability of the system. The bus admittance matrix can be analytically obtained or statistically estimated from the PMU data. If the current phasors and voltage phasors at all buses are obtained in real-time using optimal placement of PMUs and state estimation techniques, the bus admittance matrix can be calculated analytically. However, in cases where interconnection information is not known, bus admittance matrix can be statistically estimated using PMU measurements (Saadeh et al. 2016).

On premultiplication of Eq. (4.10) with $\begin{bmatrix} I_{\mathrm{LL}}^* & 0 \\ 0 & V_{\mathrm{GG}}^* \end{bmatrix}$ a relation as given below in Eq. (4.11) is obtained

$$\begin{bmatrix} I_{\mathrm{LL}}^* V_{\mathrm{L}} \\ S_{\mathrm{G}}^* \end{bmatrix} = \begin{bmatrix} I_{\mathrm{LL}}^* Z_{\mathrm{LL}} I_{\mathrm{L}} + I_{\mathrm{LL}}^* F_{\mathrm{LG}} V_{\mathrm{G}} \\ V_{\mathrm{GG}}^* K_{\mathrm{GL}} I_{\mathrm{L}} + V_{\mathrm{GG}}^* Y_{\mathrm{GM}} V_{\mathrm{G}} \end{bmatrix} \tag{4.11}$$

where $\boldsymbol{I}_{\mathrm{LL}}^*$ is defined as $(b–g) \times (b–g)$ diagonal matrix with elements of $\boldsymbol{I}_{\mathrm{L}}^*$ as the diagonal entry.

$\boldsymbol{V}_{\mathrm{GG}}^*$ is defined as $g \times g$ diagonal matrix with elements of $\boldsymbol{V}_{\mathrm{G}}^*$ as the diagonal entry and $S_{\mathrm{G}}^* = V_{\mathrm{GG}}^* I_{\mathrm{G}}$.

Equation (4.11) implies that

$$V_{\mathrm{G}} = Y_{\mathrm{GM}}^{-1} {V_{\mathrm{GG}}^*}^{-1} S_{\mathrm{G}}^* - Y_{\mathrm{GM}}^{-1} K_{\mathrm{GL}} I_{\mathrm{L}} \tag{4.12}$$

$$V_{\mathrm{L}} = \left[Z_{\mathrm{LL}} - F_{\mathrm{LG}} Y_{\mathrm{GM}}^{-1} K_{\mathrm{GL}}\right] I_{\mathrm{L}} + \left[F_{\mathrm{LG}} Y_{\mathrm{GM}}^{-1} {V_{\mathrm{GG}}^*}^{-1}\right] S_{\mathrm{G}}^* \tag{4.13}$$

$$V_{\mathrm{L}} = M_{\mathrm{L}} {V_{\mathrm{LL}}^*}^{-1} S_{\mathrm{L}}^* + N_{\mathrm{G}} {V_{\mathrm{GG}}^*}^{-1} S_{\mathrm{G}}^* = V_{\mathrm{L}}^{|} + V_{\mathrm{L}}^{||} \tag{4.14}$$

where
$\boldsymbol{M}_{\mathrm{L}} = \boldsymbol{Z}_{\mathrm{LL}} - \boldsymbol{F}_{\mathrm{LG}} \boldsymbol{Y}_{\mathrm{GM}}^{-1} \boldsymbol{K}_{\mathrm{GL}}$
$\boldsymbol{N}_{\mathrm{G}} = \boldsymbol{F}_{\mathrm{LG}} \boldsymbol{Y}_{\mathrm{GM}}^{-1}$.

$\boldsymbol{M}_{\mathrm{L}}$ and $\boldsymbol{N}_{\mathrm{G}}$ are complex matrices. $\boldsymbol{M}_{\mathrm{L}}$ measures the relative electrical distance between two load buses, while $\boldsymbol{N}_{\mathrm{G}}$ measures the relative electrical distance between load bus and generator bus. From Eq. (4.14), it is noted that the no-load voltage at a particular load bus is influenced by the apparent power at other load buses ($\boldsymbol{V}_{\mathrm{L}}^{|}$) and by the apparent power at generator buses ($\boldsymbol{V}_{\mathrm{L}}^{||}$). $\boldsymbol{N}_{\mathrm{G}}$ provides the coupling between $\boldsymbol{i}$th generator with kth load which is given by

$$|V_{lk}^{||}| = |N_{\mathrm{G}_{ki}}||S_{gi}|/|V_{gi}| \tag{4.15}$$

where $N_{\mathrm{G}_{ki}}$ is the (k, i)th element of $\boldsymbol{N}_{\mathrm{G}}$ matrix.

Also, $\boldsymbol{M}_{\mathrm{L}}$ provides the coupling between jth load with kth load and is given by

$$|V_{lk}^{\mathrm{l}}| = |M_{\mathrm{L}_{jk}}||S_{lj}|/|V_{lj}| \tag{4.16}$$

where $\boldsymbol{M}_{\mathrm{L}_{jk}}$ is the (j, k)th element of $\boldsymbol{M}_{\mathrm{L}}$ matrix.

Voltage-apparent power relation between ith generator and kth load is defined as

$$|S_{gi}|/\left[|V_{lk}^{\mathrm{ll}}||V_{gi}|\right] = 1/|N_{\mathrm{G}_{ki}}| \tag{4.17}$$

Similarly the voltage-apparent power relation between jth load and kth load is defined as

$$|S_{lj}/\left[|V_{lk}^{\mathrm{l}}||V_{lj}|\right] = 1/|M_{\mathrm{L}_{jk}}| \tag{4.18}$$

The higher the value of $|M_{\mathrm{L}_{jk}}|$ (or $|N_{\mathrm{G}_{ki}}|$), the lesser will be the voltage-apparent power coupling between the buses. This implies that, voltage-apparent power coupling between the buses will be less when the electrical distance between the buses is high. Thus, as shown in Eqs. (4.17) and (4.18), the relative electrical distance between buses in the power system given in $\boldsymbol{M}_{\mathrm{L}}$ and $\boldsymbol{N}_{\mathrm{G}}$ provides the coupling factor between bus voltages and apparent power. It may be noted that, the relative electrical distance measure being derived from bus admittance matrix reflects the topological proximity of the buses in the systems. This makes it suitable for partitioning large systems into voltage-apparent power coupling-based subsystems.

4.4.1 Defining Graph from Relative Electrical Distance

Let matrix $\boldsymbol{N}_{\mathrm{LG}}$ represent the coupling factor between generator buses and load buses in the system, such that (i, j)th element of the matrix $\boldsymbol{N}_{\mathrm{LG}}$ is

$$N_{\mathrm{LG}_{ij}} = \frac{1}{\left|N_{\mathrm{G}_{ij}}\right|} \tag{4.19}$$

where $N_{\mathrm{G}_{ij}}$ is the (i, j)th element of $\boldsymbol{N}_{\mathrm{G}}$.

Let $\boldsymbol{M}_{\mathrm{LL}}$ represent the coupling factor between load buses in the system, such that (i, j)th element of the matrix $\boldsymbol{M}_{\mathrm{LL}}$ is

$$M_{\mathrm{LL}_{ij}} = \frac{1}{\left|M_{\mathrm{L}_{ij}}\right|} \tag{4.20}$$

where $M_{\mathrm{L}_{ikj}}$ is the (i, j)th element of $\boldsymbol{M}_{\mathrm{L}}$.

The ith column of $\boldsymbol{N}_{\mathrm{LG}}$ matrix provides the list of loads that are highly coupled to ith generator bus. Mean cut is applied to the ith column of $\mathbf{N}_{\mathrm{LG}}$ matrix, and the set of loads with coupling above the mean value is grouped under the ith generator bus. The

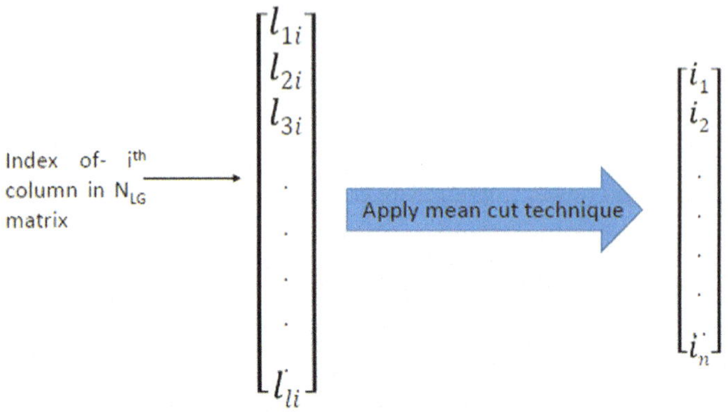

Fig. 4.1 Sorting and mean cut technique on ith column of $\boldsymbol{N}_{\mathrm{LG}}$ matrix

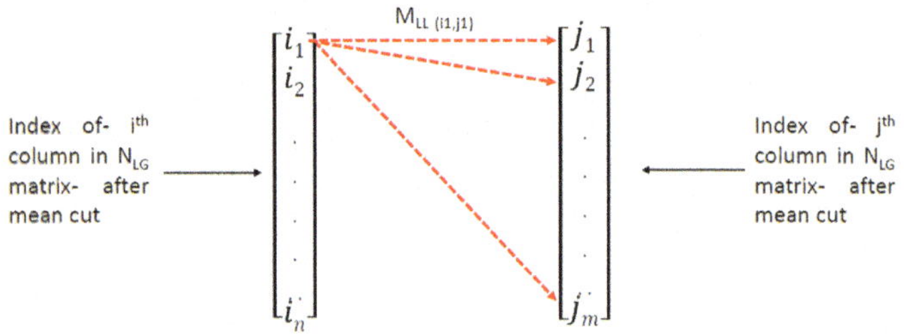

Fig. 4.2 Coupling between ith generator-loads set and jth generator-loads set

set of load buses that are highly coupled to a generator bus forms a generator-loads set. Sorting and mean cut technique is applied to each column of $\boldsymbol{N}_{\mathrm{LG}}$ resulting in g number of generator-loads sets in the system. This is depicted in Fig. 4.1.

It may be noted that, the number of load buses above mean value under the ith generator bus may not always be same as that under other generator buses. Let n be the number of load buses above mean value under ith generator bus, and let m be the number of load buses above mean value under jth generator bus. Then the ith generator-loads set has ith generator bus and n number of loads. Similarly, jth generator-loads set has jth generator bus and m number of loads.

The coupling between ith generator-loads set and jth generator-loads set is depicted in Fig. 4.2.

From Fig. 4.2, a symmetric matrix $\boldsymbol{X}$ (of order $g \times g$) that represents the coupling between generator-loads sets is defined as

$$X = \begin{cases} X_{ij} = \max_{i=1,2\ldots n}\left[\max_{j=1,2\ldots m} M_{\mathrm{LL}_{ij}}\right] \\ X_{ii} = 0 \end{cases} \tag{4.21}$$

The symmetric matrix $\boldsymbol{X}$ can be related to a complete weighted graph $\boldsymbol{G}(\boldsymbol{V}, \boldsymbol{E})$ where V is the set of vertices and E is the set of edges connecting the vertices in the graph. In this case, the set of vertices is given by g number of generator-loads sets and weight of the edges in set E is defined by the elements of matrix $\boldsymbol{X}$. The advantage of defining $\boldsymbol{X}$ matrix is that, the dimension of the large system has been reduced to the number of generators in the system, which is generally much lesser than the number of buses in the system. Graph partitioning techniques can be applied to this weighted graph to obtain the subsystems.

4.4.2 *Illustration of Graph Formation from Relative Electrical Distance Measure*

Consider IEEE 9-bus system with generation at nodes 1, 2 and 3. Then the relative electrical distance measures $\boldsymbol{N}_{\mathrm{LG}}$ and M_{LL} in the test system are depicted as in Fig. 4.3. From Fig. 4.3, by applying sorting and mean cut to the $\boldsymbol{N}_{\mathrm{LG}}$ matrix, three generator-load sets are defined, namely: $a = \{1, 4, 5, 6\}$; $b = \{2, 6, 8, 9\}$; and $c = \{3, 5, 7, 8\}$. Now the coupling between these three sets is determined from the $\boldsymbol{M}_{\mathrm{LL}}$ matrix. This is illustrated for set a in Fig. 4.4. In Fig. 4.4 Step 1: The coupling of each load in set a with b is determined from $\boldsymbol{M}_{\mathrm{LL}}$ matrix. For bus No. 4 in set a a factor $\boldsymbol{M}_{4b}$ is defined as the maximum value of entries in $\boldsymbol{M}_{\mathrm{LL}}$ matrix corresponding to bus No. 4 against the loads in set b. Similarly M_{5b} and M_{6b} are calculated. In Step 2:

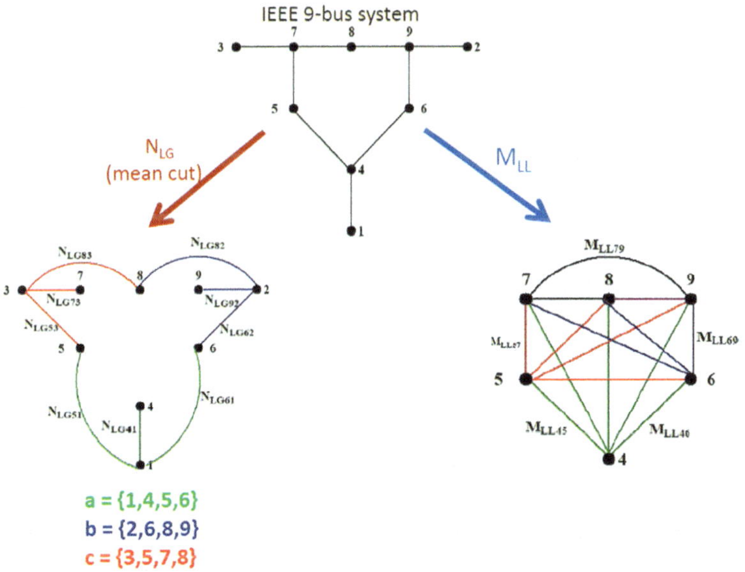

Fig. 4.3 Definition of $\boldsymbol{N}_{\mathrm{LG}}$ and $\boldsymbol{M}_{\mathrm{LL}}$ in IEEE 9-bus system

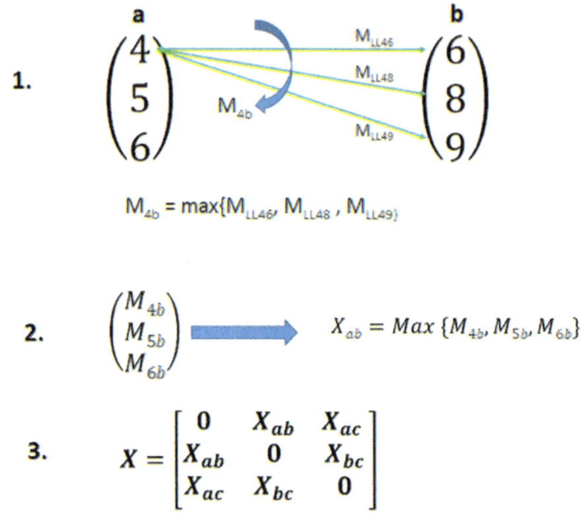

Fig. 4.4 Definition of X in IEEE 9-bus system

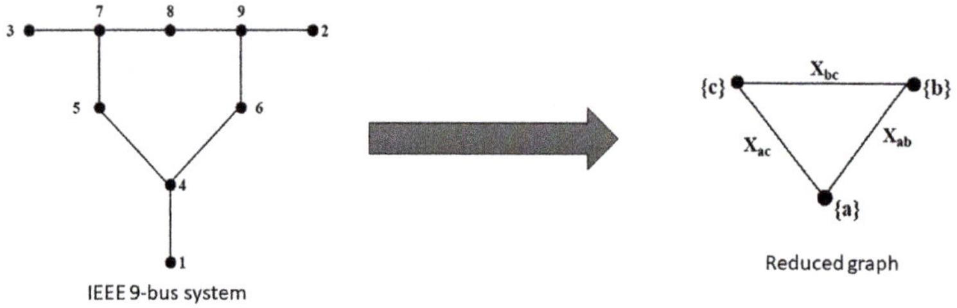

Fig. 4.5 Graph representation of IEEE 9-bus system

Coupling between sets a and b is defined as the maximum value of (M_{4b}, M_{5b}, M_{6b}). In Step 3: On performing similar operations between a–c and b–c, the symmetric matrix $\boldsymbol{X}$ can be defined. This results in a reduced graph as shown in Fig. 4.5.

Thus it can be observed from Fig. 4.5 that the large system reduces to small graph with count of vertices equal to the number of generators in the system.

4.5 Voltage-Apparent Power Coupled Areas

From the relative electrical distance measure a symmetric matrix $\boldsymbol{X}$ was defined in the above sections. This converted the large system into a reduced graph. The definition of subsystems in the large system has reduced to the problem of partitioning the reduced graph. The objective of graph partitioning problem is to minimise the total cut, and this graph partitioning is an NP-complete problem. Exact algorithms that solve the

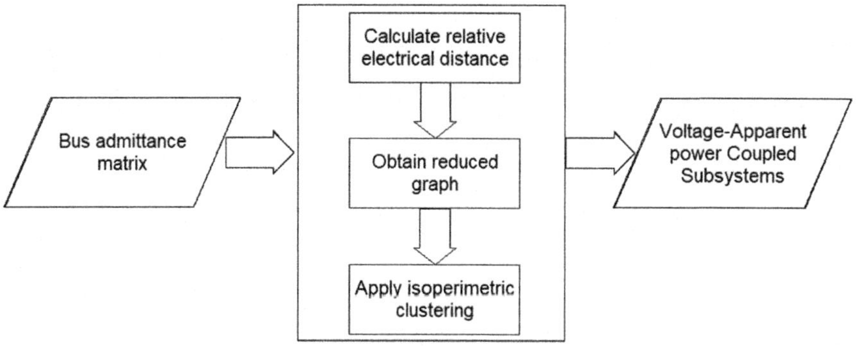

Fig. 4.6 Defining voltage-apparent power coupled areas using partitioning technique

optimisation are based on branch and bound framework which have immense running time as the size of the graph increases (Beineke et al. 2004). Many graph partitioning techniques have been proposed in literature like knapsack, max-flow min-cut (Korte et al. 2012), spectral clustering (Sánchez-García et al. 2014; Jiang et al. 2016) and isoperimetric clustering (Grady and Schwartz 2006).

Spectral decomposition techniques are applied to the Laplace matrix defined for a graph for partitioning. The advantage of applying spectral decomposition-based techniques is that Cheeger inequality (Chung and Graham 1997) provides a bound for the quality of the optimal cut. Spectral clustering techniques are based on the evaluation of eigenvalues and eigenvectors of the Laplacian of the graph. Isoperimetric clustering technique is based on the isoperimetric constant of the graph that results in solving a set of linear simultaneous equations. Among various graph clustering techniques available in the literature, isoperimetric clustering technique is found to be faster than spectral clustering techniques. This is because isoperimetric clustering technique is based on solving a system of linear equations (Grady and Schwartz 2006). Thus when compared with spectral clustering technique, isoperimetric clustering has improved speed and numerical stability. Hence isoperimetric clustering technique is applied in this work to solve the graph partitioning problem. A schematic representation of the technique to partition the system into voltage-apparent power coupled areas, proposed in Jay and Swarup (2019), is shown in Fig. 4.6.

From the bus admittance matrix the relative electrical distance measures are calculated. From the relative electrical distance measure a reduced graph that represents the large system is then derived on to which isoperimetric clustering-based network partitioning technique is applied in this work.

4.5.1 Laplacian of a Graph and Isoperimetric Clustering

The adjacency matrix $\boldsymbol{A}(\boldsymbol{G})$ of a weighted graph $\boldsymbol{G}(\boldsymbol{V}, \boldsymbol{E})$ is defined as (Beineke et al. 2004)

$$A(G) = \begin{cases} a_{vw} > 0 & \text{if} \quad vw \in E \\ a_{vw} = 0 & \text{if} \quad vw \notin E \quad \text{and} \quad a_{vw} = a_{wv} \end{cases} \tag{4.22}$$

The matrix $\boldsymbol{D}(\boldsymbol{G})$ is defined as

$$D(G) = \text{diag}\left(\deg_v : v \in V\right) \tag{4.23}$$

where

$$d = \deg_v = \sum_{w \in V} a_{vw} \tag{4.24}$$

$\boldsymbol{d}$ is called the degree vector of the graph. The Laplacian of a graph $\boldsymbol{L}$ is a positive semi definite matrix and is defined as

$$L = D(G) - A(G) \tag{4.25}$$

The isoperimetric number of graph G (Chung and Graham 1997) (h_G) cut into G_1 and $G_2 = G - G_1$ is given by

$$h_\text{G} = \frac{\text{Cut}(G_1, G_2)}{\min\left(\text{Mass}(G_1), \text{Mass}(G_2)\right)} \tag{4.26}$$

where
$\text{Cut}(G_1, G_2)$ is the sum of weights of edges between G_1 and G_2.
$\text{Mass}(G_1)$ is the sum of weights of edges incident to each vertex of the cluster G_1.
$\text{Mass}(G_2)$ is the sum of weights of edges incident to each vertex of the cluster G_2.

A well-partitioned graph implies that isoperimetric number of that graph is minimum. Hence graph partitioning problem reduces to minimising the isoperimetric number. With an indicator vector $\boldsymbol{x}$, the minimisation of cut translates to minimisation of $x^\text{T}\boldsymbol{L}x$ (Chung and Graham 1997), such that volume of each subgraph (i.e. $x^\text{T}d$) is a constant (K). The Lagrangian of the optimisation problem is

$$Q(x) = x^\text{T}Lx - \Lambda\left(x^\text{T}d - K\right) \tag{4.27}$$

where Λ is the Lagrange multiplier.

The optimisation problem thus reduces to solving a system of linear equations derived from the Lagrangian, i.e.

$$Lx = d \tag{4.28}$$

If $\boldsymbol{L}$ is singular, the node with the maximum degree is considered as the ground node (gn) (Grady and Schwartz 2006; Rege et al. 2008). The ground node is deleted from $\boldsymbol{L}$ and $\boldsymbol{d}$, with $\boldsymbol{x}(\mathbf{gn}) = 0$, resulting in another set of linear equations as

$$L_0 x_0 = d_0 \tag{4.29}$$

The solution $\boldsymbol{x}_0$ assigns real values to each node which can be then clustered using K-means or other vector clustering techniques. In the weighted graph $\boldsymbol{G}(\boldsymbol{V}, \boldsymbol{E})$ with g number of vertices, the adjacency matrix of the graph is same as the $\boldsymbol{X}$ matrix defined in Eq. (4.21), from which $D(G)$ of the graph is derived. Hence the Laplacian of the graph $\boldsymbol{G}(\boldsymbol{V}, \boldsymbol{E})$, $\boldsymbol{L}$ is defined as $\boldsymbol{L} = \boldsymbol{D}(\boldsymbol{G}) - \boldsymbol{X}$.

Since the graph is complete, any node can be considered as the ground node for applying isoperimetric clustering. For convenience, the first node is considered as the ground node. On solving Eq. (4.29), x_0 is obtained which assigns real values to each node in the graph. K-means clustering technique is applied to $\boldsymbol{x}_0$ wherein the optimal number of clusters is obtained through Elbow method (Kodinariya and Makwana 2013).

4.5.2 Algorithm of Isoperimetric Clustering-Based Network Partitioning

The concept of relative electrical distance and the algorithm for isoperimetric clustering of the system into voltage-apparent power coupled areas have been discussed in the preceding sections. The detailed flowchart of the partitioning algorithm is provided in Fig. 4.7.

The major steps involved in the partitioning of the system into voltage-apparent power coupled areas are summarised below:

1. Calculate $\boldsymbol{N}_{\mathrm{LG}}$ and $\boldsymbol{M}_{\mathrm{LL}}$ matrices from the bus admittance matrix.
2. Sort columns of $\boldsymbol{N}_{\mathrm{LG}}$ in descending order and assign loads above mean value of the column of the matrix to the corresponding generator to form generator-loads sets.
3. Find the coupling between such generator-loads sets, to form $\boldsymbol{X}$ in Eq. (4.21).
4. Obtain the Laplacian matrix $\boldsymbol{L}$ and degree vector $\boldsymbol{d}$.
5. Apply isoperimetric clustering technique to obtain $\boldsymbol{x}_0$ as in Eq. (4.29).
6. Find optimal number of clusters in $\boldsymbol{x}_0$ using Elbow method.
7. Apply K-means clustering to $\boldsymbol{x}_0$ to obtain voltage-apparent power coupled areas in the system.

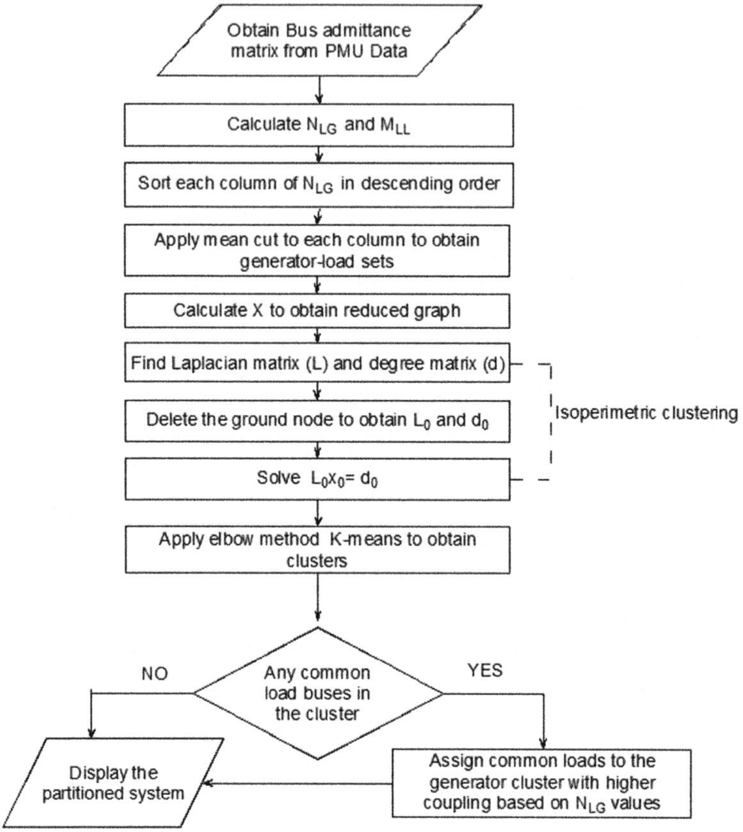

Fig. 4.7 Flowchart for partitioning system to voltage-apparent power coupled areas

4.6 Summary and Takeaways

In this chapter, network partitioning techniques were defined which is necessary to consider the localised nature of reactive power. This is essential for determining the reactive power requirement in the network. Network partitioning helps in effective reactive power management and design of reactive power market mechanism. In this chapter two modes of network partitioning are described—voltage control areas and voltage-apparent power coupled areas. A relative electric distance was described in this chapter that provides information about the coupling of voltage and apparent power requirement in the system. This measure is suitable to consider the inherent coupling between active and reactive power in real-time market operations. Since calculation of the Jacobian matrix is not involved in the relative electrical distance measure, the computation of the voltage-apparent power coupling factor is much faster. Also, isoperimetric clustering-based algorithm for partitioning the system into voltage-apparent power coupled areas is found to be adaptive to operating condition that reflects the change in apparent power requirement in the system.

References

Abou Daher N, Mougharbel I, Saad M, Kanaan HY (2013) Comparative study of partitioning methods used for secondary voltage control in distributed power networks. In: 2013 IEEE international conference on smart energy grid engineering (SEGE). IEEE, pp 1–7

Bahmanifirouzi B, Farjah E, Niknam T, Farsani EA (2012) A new hybrid HBMO-SFLA algorithm for multi-objective distribution feeder reconfiguration problem considering distributed generator units. Iran J Sci Technol Trans Electr Eng 36(E1):51

Beineke LW, Wilson RJ, Cameron PJ et al (2004) Topics in algebraic graph theory, vol 102. Cambridge University Press

Chung FR, Graham FC (1997) Spectral graph theory, vol 92. American Mathematical Society

Grady L, Schwartz EL (2006) Isoperimetric partitioning: a new algorithm for graph partitioning. SIAM J Sci Comput 27(6):1844–1866

Jay D, Swarup KS (2019) Isoperimetric clustering-based network partitioning algorithm for voltage-apparent power coupled areas. IET Gener Transm Distrib 13(22):5109–5116

Jiang T, Bai L, Jia H, Li F (2016) Spectral clustering-based partitioning of volt/VAR control areas in bulk power systems. IET Gener Transm Distrib 11(5):1126–1133

Jiang T, Bai L, Jia H, Li F (2017) Spectral clustering-based partitioning of volt/VAR control areas in bulk power systems. IET Gener Transm Distrib 11(5):1126–1133

Kodinariya TM, Makwana PR (2013) Review on determining number of cluster in K-means clustering. Int J 1(6):90–95

Korte B, Vygen J, Korte B, Vygen J (2012) Combinatorial optimization, vol 2. Springer

Mehrjerdi H, Lefebvre S, Saad M, Asber D (2012) A decentralized control of partitioned power networks for voltage regulation and prevention against disturbance propagation. IEEE Trans Power Syst 28(2):1461–1469

Rege M, Dong M, Fotouhi F (2008) Bipartite isoperimetric graph partitioning for data co-clustering. Data Min Knowl Discov 16(3):276–312

Saadeh M, Alsarray M, McCann R (2016) Estimation of the bus admittance matrix for transmission systems from synchrophasor data. In: 2016 IEEE/PES transmission and distribution conference and exposition (T&D). IEEE, pp 1–5

Sánchez-García RJ, Fennelly M, Norris S, Wright N, Niblo G, Brodzki J, Bialek JW (2014) Hierarchical spectral clustering of power grids. IEEE Trans Power Syst 29(5):2229–2237

Visakha K, Thukaram D, Jenkins L (2004) Transmission charges of power contracts based on relative electrical distances in open access. Electr Power Syst Res 70(2):153–161

Zhong J, Nobile E, Bose A, Bhattacharya K (2004) Localized reactive power markets using the concept of voltage control areas. IEEE Trans Power Syst 19(3):1555–1561

Chapter 5
Value-Based Reactive Power Market in Local Market Areas

Abstract This chapter introduces the value-based real-time reactive power market that considers the value of reactive power components in the system—the load serving component, the voltage support component and the reactive power reserve component.

Keywords Value-based reactive power market · Voltage-apparent power coupled areas · Value function · Locational marginal price

5.1 Introduction

The pricing mechanism for reactive power involves minimising a cost function related to the operation cost and the lost opportunity cost of the reactive power source. Nodal pricing scheme is considered technically efficient for real-time/spot markets, which is determined using Lagrangian dual variables of AC Optimal Power Flow (AC-OPF) based on system constraints. Market participants submit their bids to an Independent System Operator (ISO) that determines suitable price signals by defining local market areas. The price signals should reflect the requirement of reactive power in the network to provide sufficient voltage support. However, the cost function of reactive power depends on its location in the network and operating condition of the system. The cost of reactive power is not directly related to fuel costs.

In this chapter, a value-based reactive power market pricing scheme is proposed considering local market areas in the system. A value function for reactive power is defined such that the reactive power requirement in the system is assessed locally considering voltage support and reactive power reserve constraints. Based on the value function, locational marginal prices are determined in each local market area. Local market area definition based on voltage control area (VCA) and voltage-apparent power coupled areas (VAPCA) is compared in the proposed value-based reactive power market (VBRPM).

D. Jay and K. Shanti Swarup, *Market Operation for Reactive Power Ancillary Service*, Springer Tracts in Electrical and Electronics Engineering,
https://doi.org/10.1007/978-981-99-6952-4_5

5.2 Value Function for Reactive Power

A safe and secure operation of the electrical network is achieved when there is sufficient reactive power generation for maintaining bus voltage and reactive power reserve in the system. From the system operation perspective, optimal dispatch of reactive power generation is achieved not only when loads are satisfied but also when sufficient reactive power reserve is maintained. Thus in estimating the value of reactive power required in the system, reserve and voltage support requirements are also to be considered in addition to reactive power load. An effective market for reactive power is achieved when the entire system is partitioned into subsystems. This helps in handling the localised nature of reactive power. Each subsystem/region is considered as a local reactive power market. To consider the local nature of reactive power, the value of required reactive power is estimated within each subsystem/local market.

The bidding format for generation companies (GENCOs) participating in reactive power market comprises the following components:

1. Operation cost (m_{1i} in \$/MVAr).
2. Lost opportunity cost (LOC, m_{2i} in \$/MVAr^2).

Operation cost (m_{1i}) is associated with reactive power generation by ith GENCO. Lost opportunity cost is the loss incurred by the generation company in active power market when the generators reduce their active power generation to provide required reactive power support.

ISO evaluates the requirement of reactive power in each local market area to determine locational marginal prices. Reactive power required in each subsystem comprises of two parts—reactive power generation and reactive power reserve.

$$Q^A_{\text{req}} = Q^A_{\text{g}} + Q^A_{\text{res}} \tag{5.1}$$

where
Q^A_{req} is the reactive power required in subsystem **A**, Q^A_{g} is the total reactive power generation in subsystem **A** and Q^A_{res} is the total reactive power reserve in subsystem **A**.

Reactive power generation in the subsystem is required due to reactive power loading in the subsystem. However, with zero reactive power loading in the subsystem, generators must produce/absorb reactive power to maintain bus voltage magnitude within limits. Hence, the reactive power generation from ith GENCO in subsystem **A** is composed of two components:

1. Load satisfying component.
2. Voltage support component.

$$Q_{\text{g}i} = Q_{\text{l}i} + Q_{\text{mw}i} \tag{5.2}$$

where $Q_{\text{l}i}$ is the load satisfying component of reactive power generation at ith bus. $Q_{\text{mw}i}$ is the voltage support component of reactive power generation at ith bus.

However, determining the reactive power generation required for voltage support alone at the bus in real-time is challenging. Thus upper bound for reactive power generation ($Q_{\text{mw}i}^{\max}$) required at ith bus is set as

$$Q_{\text{mw}i}^{\max} = \sum_{j \in \mathbb{N}} \frac{V\text{ref}_i^2 - V_i^2 + 1}{X_{ij}} \tag{5.3}$$

The derivation for the upper bound for reactive power generation ($Q_{\text{mw}i}^{\max}$) required at ith bus is provided in Appendix B. Reactive power generation required at ith bus in a localised market area, for maintaining bus voltages, defined as $Q_{\text{mw}i}$ is a decision variable for ISO model with upper limits set by $Q_{\text{mw}i}^{\max}$ and lower limit set by $Q_{\text{mw}i}^{\min}$ (Wang and Xu 2004) as given by

$$Q_{\text{mw}i}^{\min} = \sum_{j \in \mathbb{N}} \frac{X_{ij}\left(P_{ij}\right)^2}{V_j^2} \tag{5.4}$$

Value of voltage support component of reactive power generation ($v_{Q_{\text{mw}i}}$) is defined as

$$v_{Q_{\text{mw}i}} = m_{1i} Q_{\text{mw}i} + m_{2i}\left(Q_{\text{mw}i}^2\right) \tag{5.5}$$

In Eq. (5.5), operation cost as well lost opportunity cost is associated with voltage support component of reactive power generation. This assignment of value for voltage support component is necessary to ensure that sufficient reactive power generation is available in the system locally for a better voltage profile. Hence even when there is no reactive power demand in the system, the value of reactive power generation is evaluated considering voltage support, operation cost and lost opportunity cost.

Accordingly the value of reactive power generation from ith GENCO ($v_{Q_{gi}}$) is determined as

$$v_{Q_{gi}} = m_{1i}\left(Q_{gi} - Q_{\text{mw}i}\right) + v_{Q_{\text{mw}i}} \tag{5.6}$$

where ($Q_{gi} - Q_{\text{mw}i}$) is the load serving component of reactive power generation at ith GENCO. Value of voltage support component in Eq. (5.6) is related to net reactive power generation at ith GENCO and lost opportunity cost.

Thus, generation companies are encouraged through price signals to produce/absorb reactive power required to maintain bus voltages within permissible limits. This is necessary, especially when private companies play a major role in maintaining system parameters within the limit. Also, the reactive power capability of renewable energy sources can be utilised with this formulation. When the system operator values the voltage support component of reactive power generation, renewable energy sources are encouraged to enhance their reactive power capability. Thus a market is technically efficient when the price signals issued to generation companies encourage them to produce sufficient reactive power to maintain system parameters within the permissible limits.

Value of reactive power reserve at ith GENCO can be determined either through reserve market prices or lost opportunity cost. Without loss of generality, the lost opportunity cost is considered to determine value of reactive power reserve which is given by

$$v_{Q_{\mathrm{res}i}} = m_{2i}\left(Q_{\mathrm{g}i}^{\max} - Q_{\mathrm{g}i}\right)^2 \tag{5.7}$$

where $Q_{gi}^{\max}$ is the maximum reactive power generation possible at ith GENCO.

With this, value function for reactive power required V_{Q_i} at ith GENCO is defined as

$$V_{Q_i} = m_{1i}\left(Q_{\mathrm{g}i} - Q_{\mathrm{mw}i}\right) + m_{1i}Q_{\mathrm{mw}i} + m_{2i}\left(Q_{\mathrm{mw}i}^2\right) - m_{2i}\left(Q_{\mathrm{g}i}^{\max} - Q_{\mathrm{g}i}\right)^2 \tag{5.8}$$

Thus, the value of net reactive power required (V_Q^A) in the local market area (**A**) is defined as

$$V_Q^A = \sum_{i \in A} V_{Q_i} \tag{5.9}$$

where i is the set of generators in local market area **A**.

The proposed value function and market mechanism are depicted in Fig. 5.1.

To reduce the payment burden on consumers, and achieve optimal dispatch of reactive power, the cost of reactive power generation is minimised. However, to encourage GENCOs to maintain sufficient reserve in the system, the value of reactive

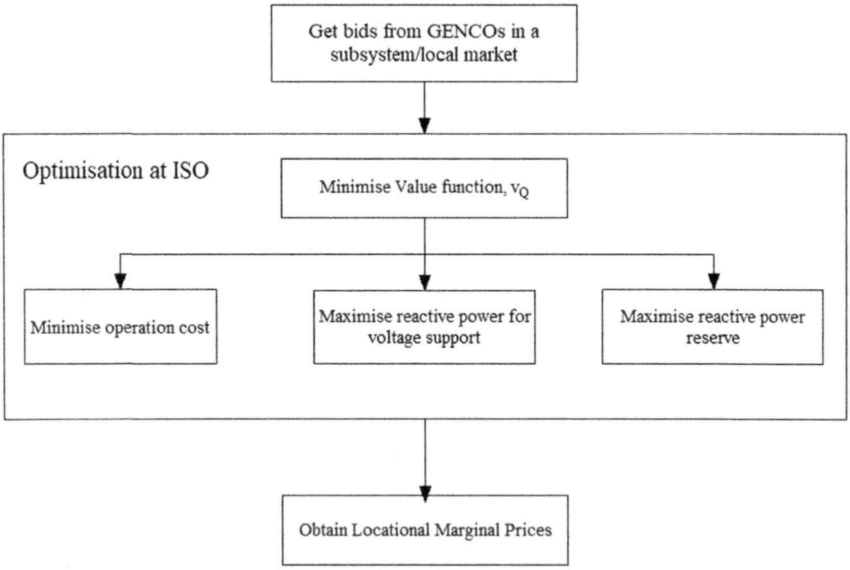

Fig. 5.1 Value-based reactive power market

power reserve is to be maximised. Reactive power market can be thus designed to minimise the proposed value function as in Eq. (5.9) in each local market area, at the system operator level. With such a definition of value function for reactive power requirement, the following objectives are achieved:

1. Minimisation of operation cost in the subsystem.
2. Maximisation of reactive power generation required for voltage support in the subsystem.
3. Maximisation of reactive power reserve in the subsystem.

5.3 Mathematical Formulation of VBRPM

In order to mitigate the effect of market power in the system, local market areas are defined in the network. This helps in considering the localised nature of reactive power. The value function defined in Eq. (5.9) is minimised in each local area by the system operator to obtain the optimal dispatch and price. The optimisation problem solved at ISO for each local area in an N bus system is defined as
Minimise

$$V_Q^A = \sum_{i \in A} m_{1i}\left(Q_{gi} - Q_{\text{mw}i}\right) + m_{1i} Q_{\text{mw}i} + m_{2i}\left(Q_{\text{mw}i}^2\right) - m_{2i}\left(Q_{gi}^{\max} - Q_{gi}\right)^2 \tag{5.10}$$

subject to the following constraints:

$$P_{gi} - P_{di} = \sum_{j \in \mathbb{N}} |V_i||V_j|Y_{ij} \cos\left(\theta_{ij} + \delta_j - \delta_i\right) \tag{5.11}$$

$$Q_{gi} - Q_{di} = -\sum_{j \in \mathbb{N}} |V_i||V_j|Y_{ij} \sin\left(\theta_{ij} + \delta_j - \delta_i\right) \tag{5.12}$$

$$\sum_{j \in \mathbb{N}} \frac{X_{ij}(P_{ij})^2}{V_j^2} \le Q_{\text{mw}i} \le \sum_{j \in \mathbb{N}} \frac{V\text{ref}_i^2 - V_i^2 + 1}{X_{ij}} \tag{5.13}$$

$$P_{gi}^{\min} \le P_{gi} \le P_{gi}^{\max} \tag{5.14}$$

$$Q_{gi}^{\min} \le Q_{gi} \le Q_{gi}^{\max} \tag{5.15}$$

$$-\ S_1^{\max} \le S_1 \le S_1^{\max} \tag{5.16}$$

$$V_i^{\min} \le V_i \le V_i^{\max} \tag{5.17}$$

Equations (5.11) and (5.12) are power balance constraints. Voltage support component, i.e. $Q_{\mathrm{mw}i}$ is considered as a decision variable in the optimisation problem with bounds given by Eq. (5.13). Real power and reactive power generation limits on generator buses are given by Eqs. (5.14) and (5.15), respectively. The transmission line constraint is included in Eq. (5.16), and bus voltages are maintained within the limit through Eq. (5.17). It is to be noted that the above problem formulation results in a non-convex nonlinear program, and thus a globally optimal solution is not guaranteed. To obtain global optimality, proper convexification techniques and the relaxation of bounds-based approach towards the problem formulation are required.

The reactive power producers are to be paid for load satisfying component and voltage support component of reactive power generation. The price signals are locational marginal prices (LMP), which are obtained from the Lagrange function ($\mathcal{L}$) of the optimisation problem given by

$$\mathrm{LMP} = \frac{\partial \mathcal{L}}{\partial Q_{\mathrm{d}i}} + \frac{\partial \mathcal{L}}{\partial Q_{\mathrm{mw}i}} \tag{5.18}$$

The above optimisation problem is applied to local market areas defined in the system. Local market areas are derived by partitioning large systems into subsystems according to system operating conditions. Hence better partitioning techniques are required for efficient localised reactive power market areas.

5.4 Case Studies

The proposed value-based reactive power market is simulated on IEEE 24-bus system as shown in Fig. 5.2. Bus No. 13 is considered as the slack bus. The network details, generation limits, etc. considered are in Grigg et al. (1996). The bid data is provided in Table 5.1. The value function-based reactive power market is solved using IPOPT solver in GAMS software. This solver is suitable to handle nonlinear programming. Three cases are considered for simulation on IEEE 24-bus system:

1. Base case analysis.
2. Line outage analysis.
3. Strategical bidding by GENCOs.

These cases are studied on IEEE 24-bus system to illustrate the following:

1. Less total voltage deviation and adequate reserve in the system.
2. Mitigating price volatility under line outage cases.
3. Effective identification of abuse of market power and reduced impact of gaming on the system.

The base case is considered to demonstrate the working of the proposed reactive power market mechanism. It also serves as a benchmark to compare line outage

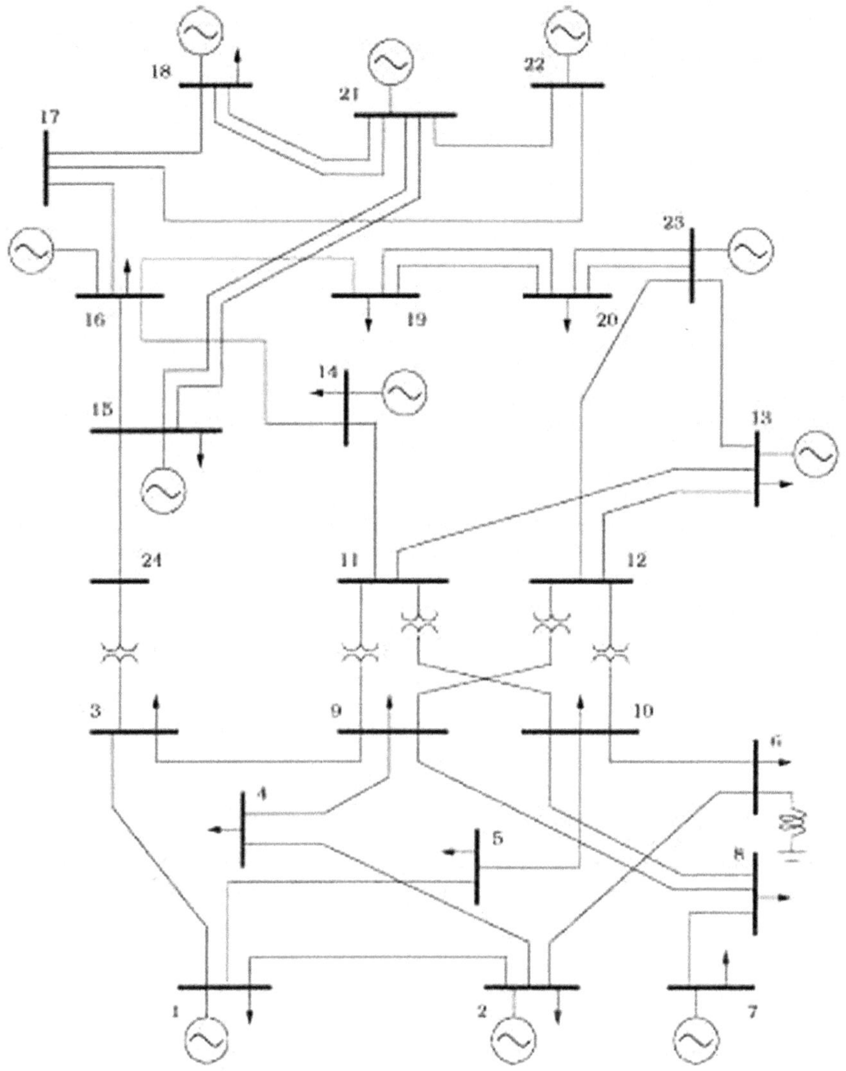

Fig. 5.2 Single line diagram of IEEE 24-bus system

cases and gaming by generation companies in reactive power market. Line outage case has been considered to demonstrate the effectiveness of the proposed market in system contingencies. The price volatility in the proposed market can be thus investigated when a line outage case is considered and compared with base case prices. Strategical bidding by generation companies is required to be simulated and analysed to check for the exercise of market power by companies in the proposed reactive power market.

Table 5.1 Bid data for IEEE 24-bus system

Generator bus number	Operation cost ($/MVAr)	LOC ($/MVAr^2)
1	0.86	0.46
2	0.68	0.39
7	0.75	0.43
13	0.60	0.50
14	0.75	0.90
15	0.73	0.38
16	0.50	0.30
18	0.85	0.48
21	0.75	0.41
22	0.48	0.38
23	0.89	0.55

Table 5.2 Local market areas defined for base case in IEEE 24-bus system

Method	No. of local market areas	GENCOs in area
VCA	3	Area 1: 1, 2, 7 Area 2: 13, 14, 23 Area 3: 15, 16, 18, 21, 22
VAPCA	2	Area 1: 1, 2 Area 2: 7, 13, 14, 15, 16, 18, 21, 22, 23

Voltage control area (VCA) based and voltage-apparent power coupled area (VAPCA) based local markets are analysed and compared in all the above cases. This provides insights regarding the importance of determining suitable partitioning techniques for defining local market areas in reactive power markets.

5.4.1 Case 1: Base Case

Base case of IEEE 24-bus system is considered to partition the system into voltage control areas (VCA) as well as voltage-apparent power coupled areas (VAPCA). The results of partitioning in both cases are tabulated in Table 5.2.

The voltage profile in the system under the proposed market is compared with that under the nodal pricing method discussed in Ahmadi and Foroud (2013). The comparison is based on the total voltage deviation (TVD) parameter defined as Ahmadi and Foroud (2013).

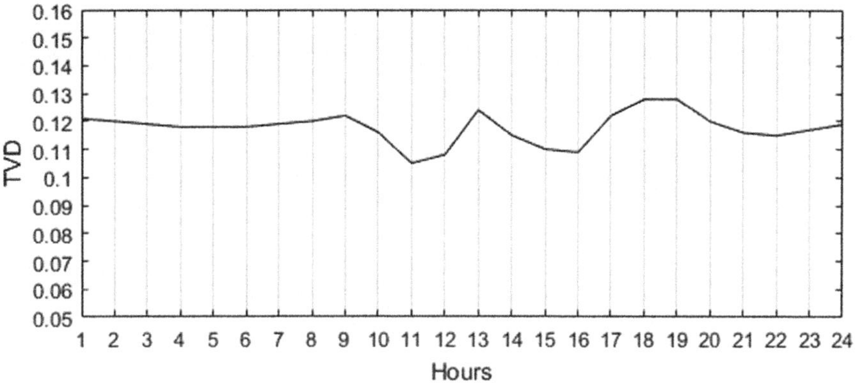

Fig. 5.3 Total voltage deviation for 24 h period

$$\text{TVD} = \sqrt{\sum_{i \in \mathbb{N}} (V_i - V\text{ref}_i)^2} \tag{5.19}$$

The total voltage deviation (TVD) in IEEE 24-bus system, when the proposed localised reactive power market settlement occurs hourly for a period of 24 h, is plotted in Fig. 5.3.

The voltages at each bus and the TVD under VCA and VAPCA based partition are tabulated in Table 5.3. Reactive power generation at each bus and payment are compared when local market areas are defined using VCA and VAPCA, which is tabulated in Table 5.4. In local areas where there is increased demand for reactive power, generations reach their maximum limit. In cases where generation plants are concentrated in a region, the reactive power is absorbed by GENCOs.

Observation and Inference

From Table 5.3, it can be observed that the proposed formulation for real-time reactive power market improves the system-wide bus voltage profile with less total voltage deviation. It can be further noticed that the total voltage deviation is much lesser with a better voltage profile in the system when voltage-apparent power coupling-based local market areas are defined (VAPCA). The improvement in TVD is 9.11% in VAPCA based partitioning technique compared with that in VCA based local market areas. Total voltage deviation is reduced in VAPCA based market areas with an increase of 158 MVar reactive power flow in the network. However, the increase in payment is 3.63% of the VCA based market model.

It can be inferred that the proposed model is able to achieve an optimal dispatch of reactive power generation by issuing price signals to generation companies, thereby obtaining a better voltage profile in the system and maintaining sufficient reserves. A better voltage profile is obtained with VAPCA based partitioning of the system, with a much less significant increase in payment. Hence system operators should decide upon their objectives in implementing reactive power support ancillary service market. When maintaining system voltage profile is of priority, VAPCA based

Table 5.3 Bus voltage and total voltage deviation-base case IEEE 24-bus system

Bus number	Nodal price based (Ahmadi and Foroud 2013)	VCA markets	VAPCA markets
1	1.035	1.023	1.037
2	1.035	1.037	1.034
3	0.963	0.977	0.981
4	0.980	0.989	0.987
5	1.010	0.987	0.995
6	0.999	0.963	0.963
7	1.025	1.036	1.037
8	0.981	0.983	0.983
9	0.970	0.985	0.983
10	1.012	0.980	0.980
11	1.003	1.011	1.010
12	1.018	0.996	0.997
13	1.030	1.037	1.037
14	0.980	1.037	1.037
15	1.014	1.003	1.011
16	1.017	1.014	1.016
17	1.039	0.985	0.994
18	1.050	0.975	0.985
19	1.023	1.014	1.004
20	1.038	1.026	1.007
21	1.050	0.981	0.991
22	1.050	0.963	0.984
23	1.050	1.004	1.013
24	0.989	0.976	0.982
TVD	0.1486	0.1155	0.1049

partitioning technique is observed to be effective. When a trade-off between payment and total voltage deviation is desired, voltage control areas-based partition will be a better choice in base case.

5.4.2 *Case 2: Line Outage Cases*

In this section, several line outage cases as mentioned below are considered in IEEE 24-bus system to comment on price volatility under contingencies:

1. Outage of the line between bus 4 and bus 9.
2. Outage of the line between bus 13 and bus 23.
3. Outage of the line between bus 16 and bus 19.

Table 5.4 MVAr generation and payment in IEEE 24-bus system—base case—VCA and VAPCA based local markets

Generator	MVAr generation		Payment	
	VCA markets	VAPCA markets	VCA markets	VAPCA markets
1	−26.95	137.51	77.46	187.19
2	192.00	32.57	547.41	62.65
7	78.28	80.13	207.97	211.62
13	201.36	201.36	883.63	905.39
14	212.17	200.00	367.14	199.57
15	145.53	215.00	135.44	588.04
16	155.00	155.00	120.29	404.98
18	−50.00	−50.00	62.37	147.27
21	−50.00	−50.00	63.67	146.45
22	−60.00	−34.01	134.82	102.66
23	−58.62	9.39	289.87	39.42
Total	738.76	896.95	2890.06	2995.25

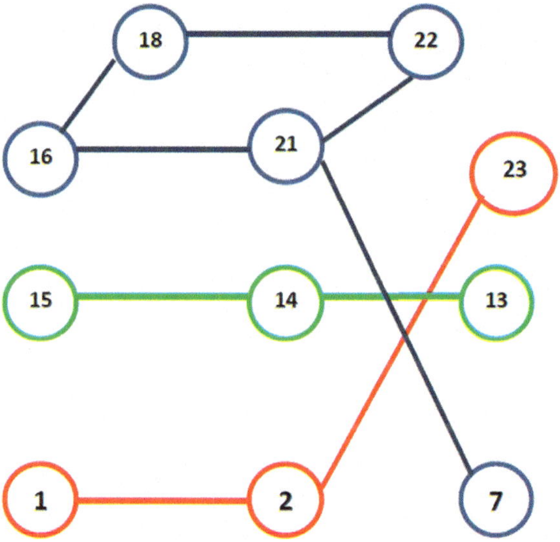

Fig. 5.4 Voltage-apparent power partitioning in IEEE 24-bus system—outage case 1, red—area 1, green—area 2, blue—area 3

In these three line outage cases, the requirement of reactive power in the system for maintaining system-wide bus voltage is different as the network configuration changes with line outage. Accordingly, the voltage-apparent power coupling between buses also changes and hence network partitioning techniques are required to be applied in all the three line outage cases resulting in different partitioned network topologies in all three cases as depicted in Figs. 5.4, 5.5 and 5.6.

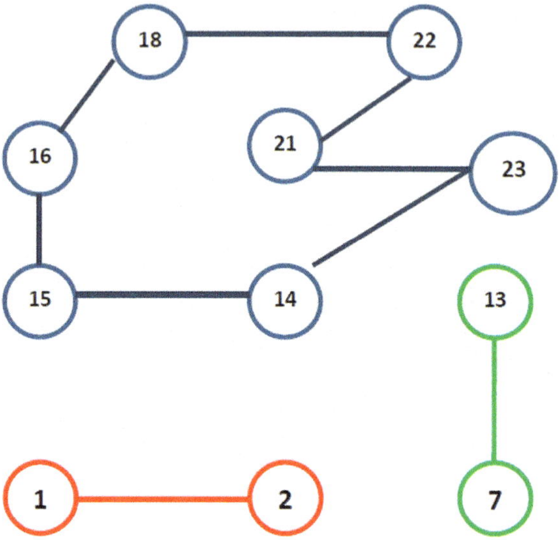

Fig. 5.5 Voltage-apparent power partitioning in IEEE 24-bus system—outage case 2, red—area 1, green—area 2, blue—area 3

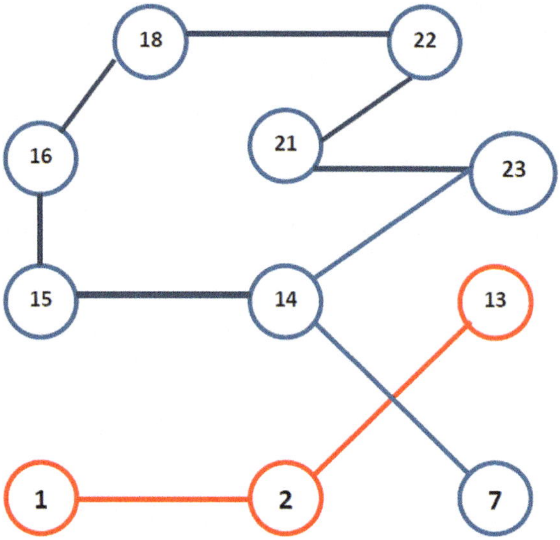

Fig. 5.6 Voltage-apparent power partitioning in IEEE 24-bus system—outage case 3, red—area 1, blue—area 2

Outage Case 1: Between Bus 4 and Bus 9

Voltage-apparent power coupling-based local market areas under outage of line between bus 4 and bus 9 are shown in Fig. 5.4. There are three local areas under VAPCA based partitioning.

Loading conditions and bid data are considered to be the same as that in the base case. This is done to check for price volatility by comparing prices in the base case with line trip cases.

Voltage profile is compared with that achieved through expected payment function (EPF)-based market model (localised). Total voltage deviation in VCA based market area model is also compared with VAPCA based market area model. The results are tabulated in Table 5.5. Net reactive power generation and payment in expected payment function (EPF) method, VCA based model and VAPCA based market model are compared in Table 5.6.

Observation and Inference

It is observed that an optimal dispatch of reactive power generation is achieved in VCA based market and VAPCA based market model as the total voltage deviation in both cases is much less than that obtained through localised expected payment method. However, the net payment is much less in localised expected payment function-based market. This is because the objective of expected payment function method is to minimise the payment and does not consider voltage support.

Table 5.5 Bus voltage and total voltage deviation—line 4–9 outage

Bus number	EPF method	VCA market areas	VAPCA market areas
1	1.037	1.024	1.037
2	1.037	1.037	1.037
3	0.976	0.976	0.977
4	0.989	0.989	0.989
5	0.994	0.988	0.995
6	0.963	0.963	0.963
7	1.037	1.037	1.031
8	0.982	0.982	0.981
9	0.983	0.984	0.980
10	0.980	0.980	0.980
11	1.006	1.010	1.008
12	1.002	0.997	1.000
13	1.037	1.037	1.032
14	1.025	1.037	1.037
15	1.029	1.003	1.003
16	1.036	1.014	1.012
17	1.033	0.985	0.984
18	1.033	0.975	0.975
19	1.026	1.014	1.012
20	1.030	1.026	1.025
21	1.037	0.981	0.981
22	1.037	0.963	0.963
23	1.037	1.010	1.008
24	0.995	0.976	0.976
TVD	0.1389	0.1168	0.1163

Table 5.6 MVAr generation and payment in IEEE 24-bus system—line 4–9 outage

Parameter	EPF method	VCA market areas	VAPCA market areas
MVAr generation	901.95	753.07	406.12
Payment in $	1613.97	2281.00	1897.65

Table 5.7 System parameters in IEEE 24-bus system—line 13–23 outage

Parameter	VCA based market areas	VAPCA based market areas
TVD	0.1157	0.1000
MVAr generation	729.3100	853.1600
Payment in $	2513.7700	3290.7300

Table 5.8 System parameters in IEEE 24-bus system—line 16–19 outage

Parameter	VCA based market area	VAPCA based market area
TVD	0.1212	0.1059
MVAr generation	864.6900	864.2000
Payment in $	2771.1290	3297.5100

The proposed market model in both VCA based partition and VAPCA based partition results in almost the same total voltage deviation. It can be further observed that the net reactive power flow is 46.07% lesser in VAPCA based model than that in VCA based model with the same total voltage deviation in the system.

Outage Case 2: Line Between Bus 13 and Bus 23

Outage of line between bus 13 and bus 23 is considered. Voltage-apparent power coupling-based local market areas under outage of line between bus 13 and bus 23 are shown in Fig. 5.5, with three market areas. Total voltage deviation, net reactive power generation and payment in VCA based model and VAPCA based market model are compared in Table 5.7. It can be observed that in both VCA and VAPCA based market models, the total voltage deviation is satisfactory. VAPCA based model offers a better voltage profile with an increase of 16.9% reactive power flow in the system when compared with VCA based model.

Outage Case 3: Line Between Bus 16 and Bus 19

The VAPCA based local market areas for outage of line between bus 16 and bus 19 case have two local market areas as shown in Fig. 5.6. Crossing occurs due to the geographic location of the bus in the system and does not depict transmission line crossing.

The system parameters obtained aftermarket settlement in both VCA based and VAPCA based methods are presented in Table 5.8.

In VAPCA based market model, an improved voltage profile is achieved compared with VCA based model with an equal net flow of reactive power in the system. This is achieved by issuing appropriate pricing signals to the generation companies to

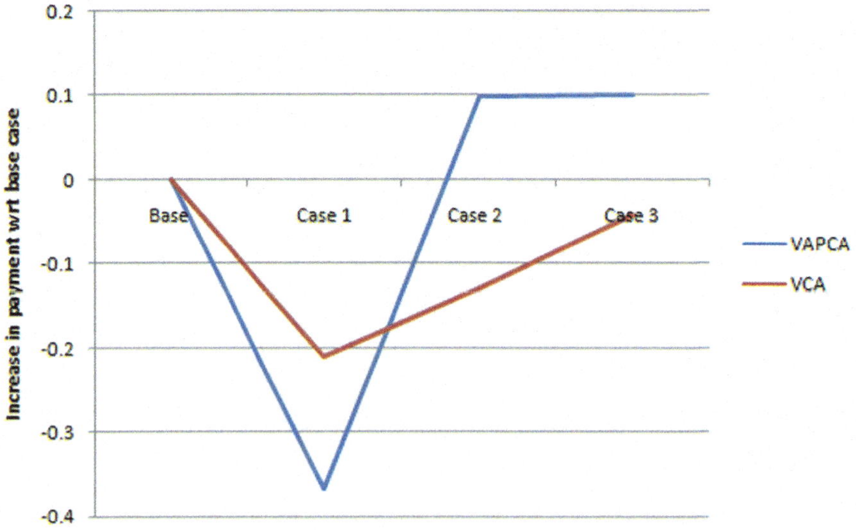

Fig. 5.7 Payment during line outage cases—comparison of VCA based partition and VAPCA based partition

maintain the system voltage profile. The increase in payment is 18.99% from VCA based model.

Observation and Inference

Price volatility has been observed in reactive power markets during contingencies like line outages. A comparative study of increment in payment on line outage cases from the base case in VAPCA based market model as well VCA based model is presented in Fig. 5.7.

Total voltage deviation in line outage cases and base case in both partitioning techniques are compared in Fig. 5.8. It is inferred that prices do not increase drastically under line outage cases. Under VCA based market model, the net payment is well controlled. However, total voltage deviation suffers. In VAPCA model, net payment increases slightly and depends on the severity of outage. However, the total voltage deviation is much lesser in all cases. Thus price volatility depends on the partitioning techniques as well as the severity of line outage. VAPCA model assures a better voltage profile in the system without a drastic increase in payment. VCA model assures control in net payment, but TVD depends on the severity of line outage.

5.4.3 Case 3: Strategical Bidding of GENCOs

Market power has been defined as the capability of participants to alter prices profitably above the competitive level. This is mostly done by submitting bids higher than actual cost, thereby increasing the price received, i.e. gaming of GENCOs. The

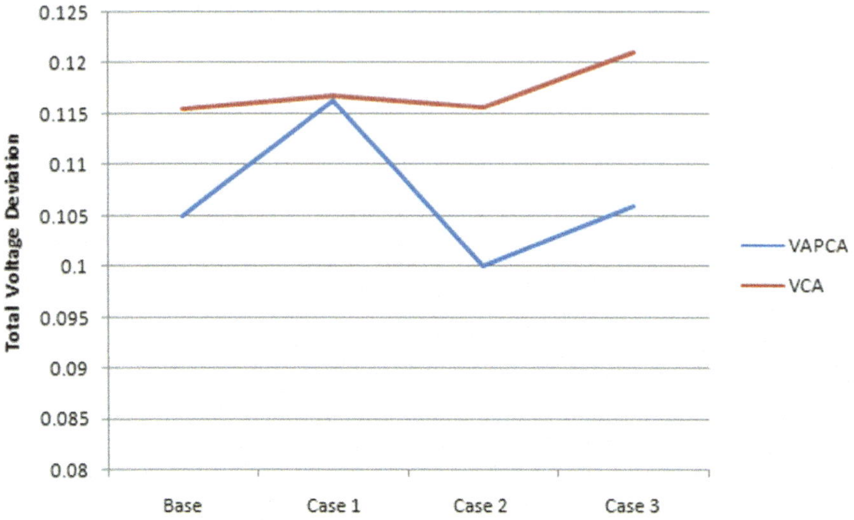

Fig. 5.8 Total voltage deviation during line outage cases—comparison of VCA based partition and VAPCA based partition

effect of market power is reduced by defining local market areas. This is simulated in VCA based as well as VAPCA based market model for IEEE 24-bus system. Four instances of gaming by GENCO at bus No. 2 are considered for simulation.

1. Scenario 1: Operation cost and lost opportunity cost bids are twice the actual cost.
2. Scenario 2: Bid for operation cost is four times the actual, bid for LOC is twice the actual lost opportunity cost.
3. Scenario 3: Bid for operation cost is five times the actual, bid for LOC is twice the actual lost opportunity cost.
4. Scenario 4: Bid for operation cost and lost opportunity cost are five times the actual cost.

Comparative study on VCA based and VAPCA based models is carried out in all the above four scenarios. Gaming results for all four scenarios are consolidated and presented in Table 5.9. Results for the most severe case of gaming, i.e. Scenario 4 in Table 5.10 under VCA model and in Table 5.11 under VAPCA model.

Observation and Inference

From Table 5.9, it is observed that gaming does not affect total voltage deviation in the system. Thus voltage profile is still maintained within satisfactory limits, with VAPCA based model giving better results. This is because the net MVAr generation schedule is not affected by gaming. On observing Table 5.10, VCA based market model results in a shift in MVAr generation schedule at each generator bus with gaming. On further investigation of Table 5.10 gaming by generation bus 2 in VCA based model affects prices of other generation companies within the local market area, i.e. Generator bus 1 and 7.

Table 5.9 Strategical bidding in IEEE 24-bus system—consolidated results

Case number	Total voltage deviation		Payment at bus 2 ($)	
	VCA	VAPCA	VCA	VAPCA
Base	0.1160	0.1049	547.40	62.65
Scenario 1	0.1160	0.1075	128.16	125.22
Scenario 2	0.1160	0.1075	171.59	169.45
Scenario 3	0.1160	0.1075	193.45	191.57
Scenario 4	0.1160	0.1075	313.98	312.97

Table 5.10 Scenario 4—strategical bidding in IEEE 24-bus system under VCA based local market areas

Generator	MVAr generation (MVAr)		Price at bus 2 ($/MVAr)	
	Base	VCA	Base	VCA
1	−26.95	136.70	2.87	1.36
2	192.00	33.42	2.85	9.61
7	78.27	78.83	2.65	2.64
13	201.35	212.16	4.38	4.39
14	212.16	145.53	1.73	1.73
15	145.53	188.59	0.93	0.93
16	155.00	155.00	0.77	0.77
18	−50.00	−50.00	1.24	1.24
21	−50.00	−50.00	1.27	1.27
22	−60.00	−60.00	2.24	2.24
23	289.87	−58.62	4.94	4.94

Table 5.11 Scenario 4—strategical bidding in IEEE 24-bus system under VAPCA based local market areas

Generator	MVAr generation (MVAr)		Price at bus 2 ($/MVAr)	
	Base	VAPCA	Base	VAPCA
1	137.50	137.59	1.36	1.36
2	32.57	32.53	1.92	9.62
7	80.13	80.13	2.64	2.64
13	201.35	201.35	4.49	4.49
14	200.00	200.00	1.00	1.00
15	215.00	215.00	2.73	2.73
16	155.00	155.00	2.61	2.61
18	−50.00	−50.00	2.94	2.94
21	−50.00	−50.00	2.92	2.92
22	−34.01	−34.01	3.01	3.01
23	9.39	9.39	4.19	4.19

On observing Table 5.11 pertaining to VAPCA based market model, MVAr generation schedule is not affected by gaming by one GENCO. This shows that the impact on scheduling due to gaming is reduced when value-based market model proposed in this work is implemented on voltage-apparent power coupled area-based partitioned market areas. In VAPCA based model, the price at bus No. 2 is alone influenced by gaming while other locational marginal prices remain the same as in the base case. The prices at other GENCOs are not affected by exercise of market power by bus No. 2 in VAPCA based market model. However, the prices at bus 2 increase as the magnification of actual cost increases. The effect is more when gaming is done on lost opportunity cost, as in case 4. This is because reserve value has also been considered at the lost opportunity cost. This situation can be avoided by considering price settlement at the reactive power reserve market to evaluate reactive power reserve in the optimisation problem.

5.5 Summary and Takeaways

Exercise of market power due to gaming by generation companies and price volatility due to system contingencies imposed challenges in implementing real-time reactive power market and pricing. These issues can be mitigated by determining the value of reactive power requirement in the system. In this chapter, value-based real-time reactive power market has been proposed, which evaluates the value of reactive power requirement in the system. The value of reactive power requirement in the system has three components—the load serving component, the voltage support component and the reactive power reserve component. The objective of value-based reactive power market mechanism is to minimise the value function of reactive power required to maintain system-wide bus voltages and sufficient reactive power reserve in the system. Simulation results on IEEE 24-bus system ensure that such a market model helps in improving the voltage profile of the system, especially when voltage-apparent power coupling-based partition is applied to identify local market areas.

References

Ahmadi H, Foroud AA (2013) A stochastic framework for reactive power procurement market, based on nodal price model. Int J Electr Power Energy Syst 49:104–113

Grigg C, Wong P, Albrecht P et al (1996) Reliability test system task force, the IEEE reliability test system-1996, a report prepared by the reliability test system task force of the application of probability methods subcommittee. IEEE Trans Power Syst 14(3):1010–1020

Wang Y, Xu W (2004) An investigation on the reactive power support service needs of power producers. IEEE Trans Power Syst 19(1):586–593

Part III
Analysis of Reactive Power Markets

Chapter 6
Game Theoretical Analysis

Abstract This chapter presents an analysis of market mechanisms through a game theoretic approach considering incentive compatibility and individual rationality.

Keywords Single-leader multi-follower game · Incentive compatibility · Individual rationality · Equilibrium · Game theory

This chapter presents the basics of game theory and market models required for market analysis. The attributes of market mechanisms are then discussed which will ensure market efficiency.

6.1 Basics of Game Theory

Game theory is the study of mathematical models of interaction between intelligent decision makers generally known as agents or players. The interaction between agents/players can be modelled generally as cooperation games or non-cooperation games. Mathematical techniques for analysing situations in which decision of a player influences the decision of other players are provided by game theory.

The three elements in a game are the set of players (N), the set of strategies (S_i) and their pay-off functions (u_i). When analysing a game, it is necessary to make clear the implications of the rules and understand the relationships between the strategies and the payoffs. Then it is necessary to determine the strategy each player will play by describing the optimal strategy given the strategy profile of other players. This describes the equilibrium strategies like strong dominance, weak dominance and very weak dominance strategies. A dominant strategy equilibrium requires that each player's choice be the best response against all possible choices of all the other players. When each player's choice is the best response to the best response strategies of other players in the game, Nash equilibrium of the game is achieved.

Given a strategic form game, the strategy profile $s^* = \{s_1^*, s_2^*, \ldots, s_n^*\}$ is said to be a pure strategy Nash equilibrium of game if,

$$u_i(s_i^*, s_{-i}^*) \geq u_i(s_i, s_{-i}^*) \quad \text{for} \quad i \in \{1, 2, \ldots n\} \tag{6.1}$$

D. Jay and K. Shanti Swarup, *Market Operation for Reactive Power Ancillary Service*, Springer Tracts in Electrical and Electronics Engineering,
https://doi.org/10.1007/978-981-99-6952-4_6

That is, each player's Nash equilibrium strategy is the best response to the Nash equilibrium strategies of the other players.

Electricity markets are studied under the category of Oligopoly markets. Oligopoly markets are those markets in which only a few firms compete, and such firms produce homogeneous or differentiated products and where barriers to entry exist that may be natural or constructed. There are three main models of oligopoly markets, namely

1 Cournot model.
2 Bertrand model.
3 Stackelberg model.

In Oligopoly market price is a function of the total quantity produced which, in turn, implies that one firm's output affects not only the price it receives for its output but the price its competitors receive as well. This creates a strategic environment where one firm's profit maximising output level is a function of their competitors' output levels.

6.1.1 Cournot Model

The oligopoly model in which each player assumes that rivals will continue producing their current output levels is called Cournot model. In Cournot model, player decides their quantity of production simultaneously. Thus each player considers other players' quantity as a fixed number.

In Cournot model, each player chooses its production quantity so as to maximise its profit. Given the rival players' quantity, the player maximises its profits by choosing the optimal quantity where marginal revenue equals to marginal cost. Each player's quantity depends on the quantity of rival players through a reaction curve. The equilibrium quantities for the players occur at the intersection of the reaction curves of each player.

Let us consider a Cournot duopoly model. For a linear demand curve,

$$P = a - b(Q_1 + Q_2) \tag{6.2}$$

The marginal revenue curve for the first player is determined by considering that the first player was a monopolist facing the residual demand curve in Eq. 6.2. To maximise profit, marginal cost is equal to marginal revenue, which is defined as the reaction curve.

$$0 = a - bQ_1 - bQ_2 \tag{6.3}$$

$$Q_1 = \frac{a}{2b} - \frac{Q_2}{2} \tag{6.4}$$

$$Q_2 = \frac{a}{2b} - \frac{Q_1}{2} \tag{6.5}$$

Equilibrium occurs at the point of intersection of the two reaction curves. For this, solve Eq. 6.4 for Eq. 6.5. Thus equilibrium occurs at

$$Q_1 = Q_2 = \frac{a}{3b} \tag{6.6}$$

On comparing the equilibrium with share monopoly, in which each player produces half of the total quantity, Cournot duopoly model produces more and charges a lower price and thus earns less profit.

6.1.2 Bertrand Model

In Bertrand model each player chooses the price simultaneously assuming that other players will continue charging their current prices. If player charges more than other players in the market, the sales may become zero. Hence it will lower prices. If player charges same as other firms, the market splits. Thus it is better to charge a small amount less than the rival to get the whole market.

6.1.3 Stackelberg Model

Oligopoly model in which one player (leader) picks the quantity before the other players (follower). The leader knows that the follower will pick output according to the Cournot reaction function in Eq. 6.5.

Residual demand facing the leader is given by

$$P = a - b\left(\frac{a}{2b} - \frac{Q_1}{2}\right) - bQ_1 \tag{6.7}$$

Solving for marginal revenue equals to marginal cost, we get equilibrium at

$$Q_1 = \frac{a}{2b} \tag{6.8}$$

$$Q_2 = \frac{a}{4b} \tag{6.9}$$

6.2 Game Theoretical Analysis of Market Models

Locational marginal pricing mechanism achieves social optimality in a perfect competitive market. However, there are growing concerns on the exercise of market power through strategical bidding and capacity withholding by market participants, especially in ancillary service markets like reactive power markets. A few theoretical works have proven that large generators can significantly raise electricity prices by withholding capacity or by bidding strategically. This results in considerable loss to Independent System Operator and affects the economic efficiency [3–5]. In Chap. 5, a single-leader multi-follower game model was discussed for reactive power pricing mechanism, where price signal was modelled as a decision variable of the Mathematical Programming with Equilibrium Constraints (MPEC). The technical efficiency of the price signal in obtaining sufficient reactive power in the network for safe and secure operation of the grid was discussed in the chapter. The existence of equilibrium in the single-leader multi-follower game model is discussed. Conditions for incentive compatibility and individual rationality are derived, which will ensure the economic efficiency of the proposed pricing mechanism. Economic efficiency of mechanism-1 is illustrated in PJM 5-bus system in the case of strategical bidding.

6.3 Existence of Equilibrium

In the MPEC formulation described for mechanism-1 in Chap. 5, the objective function is given by

$$\begin{aligned} \mathrm{UG}_{\mathrm{ISO}}(\lambda_i, Q_{\mathrm{mwi}}, Q_{\mathrm{gi}}) = \sum_{i \in \mathbb{G}} \lambda_i \cdot Q_{\mathrm{gi}} + (m_{1i} + m_{2i} \cdot Q_{\mathrm{mwi}}) \cdot Q_{\mathrm{mwi}} \cdot R_i \\ + m_{2i} \cdot (Q_{\mathrm{gi}}^{\max} - Q_{\mathrm{gi}})^2 \end{aligned} \tag{6.10}$$

The optimal value of $\mathrm{UG}_{\mathrm{ISO}}$ in Eq. (6.10) is:

$$\mathrm{UG}^*_{\mathrm{ISO}} = \inf \sum_{i \in \mathbb{G}} \lambda_i \cdot Q_{\mathrm{gi}} + C_{\mathrm{mwi}} + \mathrm{Cres}_i \tag{6.11}$$

$$\sum_{i \in \mathbb{G}} \inf[\lambda_i \cdot Q_{\mathrm{gi}} + C_{\mathrm{mwi}} + \mathrm{Cres}_i] \le \mathrm{UG}^*_{\mathrm{ISO}} \tag{6.12}$$

$$\sum_{i \in \mathbb{G}} \inf \ \mathcal{TG}_i \le \mathrm{UG}^*_{\mathrm{ISO}} \tag{6.13}$$

where

$$\begin{aligned} \mathcal{TG}_i = \lambda_i \cdot Q_{\mathrm{gi}} + (m_{1i} + m_{2i} \cdot Q_{\mathrm{mwi}}) \cdot Q_{\mathrm{mwi}} \cdot R_i \\ + m_{2i} \cdot (Q_{\mathrm{gi}}^{\max} - Q_{\mathrm{gi}})^2 \end{aligned} \tag{6.14}$$

$\lambda_i \cdot Q_{\mathrm{gi}}$ is the optimal payment that links the decision variables of leader and followers and is a positive quantity. $\mathcal{TG}_i$ can be considered as the utility function of a pseudo-leader, which issues prices signal to ith GENCO alone subject to constraints from

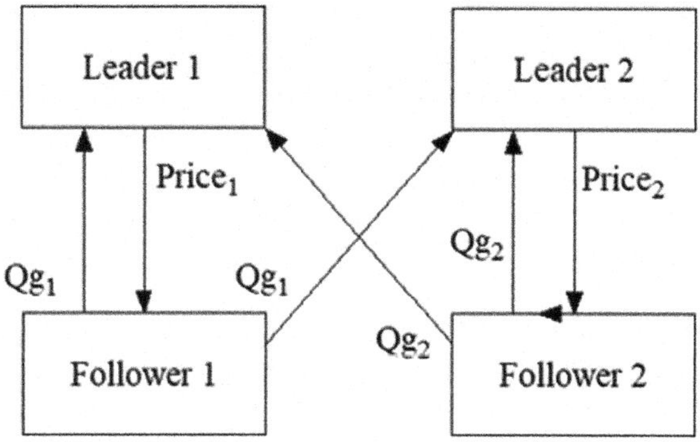

Fig. 6.1 Translation of single-leader multi-follower game to multi-leader multi-follower game model

Eqs. (??–??). This implies that each GENCO (follower) receives its price signal from a pseudo-leader whose utility function is given by $\mathcal{TG}_i$.

This translates the single-leader multi-follower game to a multi-leader multi-follower game in both the mechanisms, as depicted in Fig. 6.1.

The number of pseudo-leaders is the same as the number of followers, i.e. the number of GENCOs participating in mechanism-1 and the number of aggregators in mechanism-2. Thus the single-leader multi-follower game can be viewed as a multi-leader multi-follower game (MLMF) with shared constraints. The pseudo-leaders, each with $\mathcal{TG}_i$ as utility function, form the multi-leader set and GENCOs form the multi-follower set in mechanism-1. The existence of equilibrium/equilibria of the proposed single-leader multi-follower game can be thus proven using potential function game representation of MLMF game (Kulkarni and Shanbhag 2014).

Lemma 1 *$\pi(x)$ is defined as a potential function for a function f(x), $x \in X$, if for all $x' \in X$, $\pi(x) - \pi(x') = f(x) - f(x')$.*

In $\mathcal{TG}_i$, the utility function of the ith pseudo-leader that issues price signal to ith GENCO in the MLMF game, the bilinear term $\lambda_i \cdot Q_{\text{gi}}$ is approximated with first-order Taylor series expansion around (λ_i,Q_{gi}) values in the previous time slot of market operation. Let 'l' be value of λ_i and 'q' be value of Q_{gi} in the previous market clearing. Thus on applying Taylor series expansion around the point (l, q) to the optimal payment term $\lambda_i \cdot Q_{\text{gi}}$ in Eq. (6.14),

$$\mathcal{TG}_i = q \cdot \lambda_i + l \cdot Q_{\text{gi}} - l \cdot q + m_{2i} \cdot R_i \cdot Q^2_{\text{mwi}} + m_{2i} \cdot (Q^{\max}_{\text{gi}})^2 + m_{2i} \cdot Q^2_{\text{gi}} \tag{6.15}$$

It may be noted that GENCOs may either generate or absorb reactive power, and thus the term $l \cdot Q_{\text{gi}}$ may not always be positive, thereby violating the condition of optimal payment term to be positive. This is rectified by squaring the terms, and without loss of generality, the approximate utility function of ith pseudo-leader is defined as,

$$\mathcal{TG}_i = \lambda_i^2 + m_{2i} \cdot R_i \cdot Q^2_{\text{mwi}} + m_{2i} \cdot (Q^{\max}_{\text{gi}})^2 + [m_{2i} + 1] \cdot Q^2_{\text{gi}} - l \cdot q \tag{6.16}$$

By Lemma 1, the potential function π_{gi} for the utility function $\mathcal{TG}_i$ is defined as,

$$\pi_{\text{gi}} = \lambda_i^2 + m_{2i} \cdot R_i \cdot Q_{\text{mwi}}^2 + [m_{2i} + 1] \cdot Q_{\text{gi}}^2 \tag{6.17}$$

Similarly the potential function π_a for the utility function $\mathcal{TA}_a$ is defined as,

$$\pi_a = \frac{\lambda_i^2}{b_i^2} + p_{2i} \cdot R_i \cdot Q_{\text{va}}^2 + [p_{2i} + 1] \cdot Q_a^2 \tag{6.18}$$

Theorem 1 (Existence of equilibrium): *Suppose the set of fixed points in the feasible region of multi-leader multi-follower game-based reactive power market is non-empty. If utility function of each leader is continuous, then the multi-leader multi-follower game with potential function π has an equilibrium if*

1 The potential function π is coercive.
2 Or the constraint space is compact.

i.e. there exists a minimiser for the potential game defined by the potential function π, which defines the equilibrium of multi-leader multi-follower game.

Proof To prove the existence of equilibrium in multi-leader multi-follower games with shared constraints, it is required to derive a potential function π which is coercive. Thus the minimiser for potential game exists, and by the definition of potential function, an equilibrium for the multi-leader multi-follower game exists (Kulkarni and Shanbhag 2014).

The coercivity of potential functions π_{gi} and π_a in Eqs. (6.17) and (6.18), respectively, is now straightforward. Coercivity is a strong form of convexity, and thus a minimiser to the potential game representation of the multi-leader multi-follower game exists. By the definition of potential function, it can also be understood that minimiser of the potential function represents the equilibrium of the multi-leader multi-follower game with shared constraints. This proves the existence of equilibrium in the single-leader multi-follower game which was translated as a multi-leader multi-follower game with shared constraints.

Regarding the uniqueness of the equilibrium point, for a potential game to achieve a global minimiser, the following conditions are to be satisfied

1. Potential function is convex.
2. The set of fixed points within the feasible region of the multi-leader multi-follower game is also convex.

However, the non-convexity in the ISO system constraints can guarantee only local minima. The non-convexity can be handled using suitable convexification techniques, and with feasible set compactness it is possible to achieve global minimiser for the potential game, thereby arriving at a unique equilibrium in the multi-leader multi-follower game. Hence, the existence of equilibrium is proven; however, uniqueness cannot be guaranteed due to non-convexity. Convexification techniques help in arriving at approximate solutions.

6.4 Incentive Compatibility and Individual Rationality

A pricing mechanism is said to be incentive compatible when the market participants achieve their best outcome when it declares its actual type in the auction/bidding process. Thus an incentive compatible market mechanism induces a truth-telling behaviour among participants, thereby mitigating market power. Such price mechanisms are termed as direct revelation mechanisms. Market mechanisms that preserve incentive compatibility are thus economically efficient. In proposed reactive power market, pricc signal issued by ISO to each GENCO depends on marginal cost function bids submitted by GENCOs.

Consider a market with tariff mechanism as $t(\Theta)$ and $x(\Theta)$ as the decision variable where Θ is the type of the participant. Then allocation $y = (x(\Theta), t(\Theta))$ is said to be direct revelation mechanism if agents in the market simultaneously and truthfully announce their types. Thus such a mechanism that induces truth-telling behaviour among the agents (GENCOs) to declare their actual type (marginal cost function) is said to be incentive compatible.

Let $x(\Theta)$ be the decision function of participant with type Θ in the market. Let the tariff be $(t\Theta)$. Then an allocation $y = (x(\Theta), t(\Theta))$ is said to be direct revelation mechanism if $x(\Theta)$ is implementable.

Lemma 2 *A decision function $x(\Theta)$ is implementable if and only if it is monotone (non-decreasing). Also, if $x(\Theta)$ is implementable there exists a tariff mechanism $t(\Theta)$ such that allocation $y = (x(\Theta), t(\Theta))$ is incentive compatible (Fudenberg and Jean 1991).*

The following conditions are considered as an equivalent to monotonicity condition on the decision function $x(\Theta)$ (Fudenberg and Jean 1991):

1. Sorting condition or Spence-Mirrlees condition, which is expressed as

$$\frac{\partial}{\partial \Theta}\left(\frac{\partial u/\partial x}{\partial u/\partial t}\right) \geq 0 \tag{6.19}$$

   where u is the utility function of the agent with decision function x.
2. There exists a p and q such that,

$$\left|\frac{\partial u/\partial x}{\partial u/\partial t}\right| \leq p + q\,|t| \tag{6.20}$$

   uniformly over x, t, Θ

There are two strategies by which a participant exercises market power. These strategies are:

1. Bidding prices higher than the actual cost (financial withholding).
2. Withholding actual capacity while bidding (capacity withholding).

Thus participants are of two types—financial withholding type (Θ_1) and quantity withholding type (Θ_2). Incentive compatibility under these two strategies is to be evaluated to ensure that the market is economically efficient.

Incentive compatibility with financial withholding

Theorem 2 *Let λ_i be the price signal issued to ith GENCO with marginal cost function Θ_1 under the single-leader multi-follower game-based reactive power market model. The mechanism is said to be incentive compatible if*

1. $\lambda_i = a.\Theta_1 + b$
2. $a > 1$
3. $b > (1 - a).\Theta_1$.

Proof By Lemma 2, a monotone non-decreasing decision function is implementable. The equivalence to monotonicity of a function is given by Eqs. (6.19) and (6.20). Hence decision function ($x(\Theta_1)$) that satisfies Eqs. (6.19) and (6.20) is implementable. Thus Lemma 2 implies that, the allocation ($x(\Theta_1)$, $t(\Theta_1)$) is incentive compatible.

In the proposed mechanism, the price signal issued to a GENCO with marginal cost function Θ_1 is tariff, and corresponding optimal reactive power generation is the decision function. Then the utility function for the GENCO with marginal cost function Θ_1 is defined as

$$u = t(\Theta_1).x(\Theta_1) - \int_0^{x(\Theta_1)} \Theta_1(x)\mathrm{d}x \tag{6.21}$$

Thus $x(\Theta_1)$ be implementable by satisfying the conditions provided in Eqs. (6.19) and (6.20) which is achieved with $t(\Theta_1) = a.\Theta_1 + b$ and the following:

1. $a > 1$
2. $b > (1 - a).\Theta_1$.

By imposing the above conditions on the price signal issued by ISO, incentive compatibility is achieved in the proposed reactive power market mechanism under equilibrium.

From the conditions derived in Theorem 2, under equilibrium it can be noticed that the price received

$$\lambda = a.\Theta_1 + b > a.\Theta_1 + (1 - a).\Theta_1 > \Theta_1 \tag{6.22}$$

The mechanism induces a truth-telling behaviour among participants under equilibrium. Moreover, at the same time, they are assured a non-negative profit. Thus individual rationality is preserved. This thus helps in mitigating the exercise of market power in the proposed reactive power market.

Table 6.1 Reactive power cost for generators in PJM 5-bus system

Bus	Operation cost ($/100 MVAr)	Lost opportunity cost ($/100 MVAr^2)
1	0.86	0.46
1	0.86	0.46
3	0.68	0.39
4	0.75	0.43
5	0.60	0.50

6.5 Illustrative Examples

Consider PJM 5-bus system, as shown in Fig. 6.2. The cost function is provided in Table 6.1. The proposed market model is compared with the locational marginal price-based model. LMP model considered in this work is based on AC-OPF formulation with ISO utility function as objective function. To demonstrate economic efficiency of the proposed system, we consider the case when the generator at bus 1 reports a cost function different from actual (magnifies their actual cost of production by factors ranging from 1 to 6). The remaining GENCOs report their actual cost.

The actual operation cost of reactive power generation at bus 1 is 0.86$/100 MVAr. When it bid its actual operation cost and lost opportunity cost, the price received

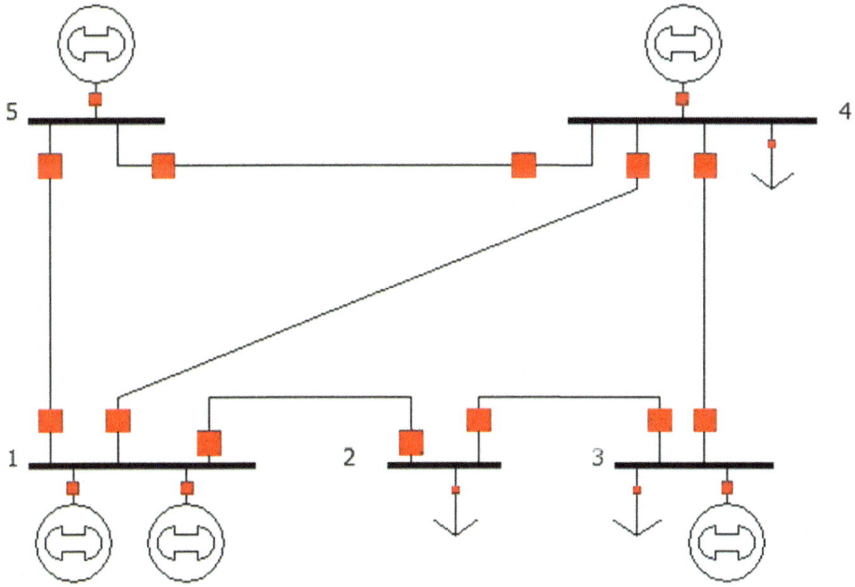

Fig. 6.2 PJM 5-bus system

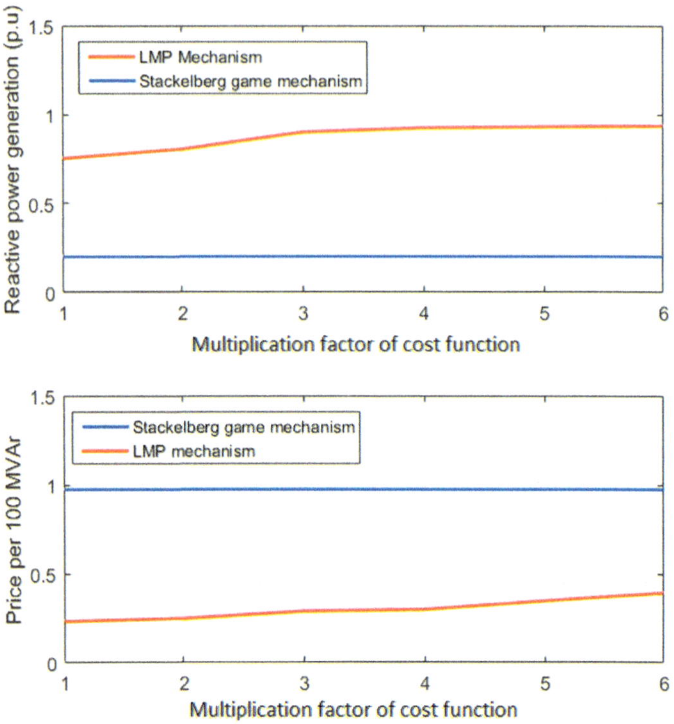

Fig. 6.3 Price per 100 MVAr and p.u reactive power generation at bus 1 when cost function was magnified during bid

through locational marginal pricing (LMP) mechanism is 0.234$/100 MVAr, which is much lesser than its actual operation cost. Thus GENCOs tend to magnify their actual cost while bidding. When GENCO at bus 1 bids an operation cost of 5.16$/100 MVAr, which is six times the actual operation cost, under LMP mechanism the price received is only 0.396$/100 MVAr,thereby GENCO further magnifying their actual cost. Whereas in the proposed single-leader multi-follower game model, the price received by GENCO bus 1 is 0.976$/100 MVAr. The same price is received even if GENCO magnifies its actual cost while bidding. Hence generation company does not gain additional benefits by bidding false cost functions. The price received under single-leader multi-follower game model is 13.48% higher than the operation cost. Thus by bidding actual operation cost, GENCOs receive a non-negative profit, thereby preserving individual rationality. The price received per 100 MVAr and the reactive power generation under both mechanisms is compared in Fig. 6.3.

The total cost of reactive power generation (quadratic cost function) at bus 1 under LMP mechanism is compared with game model. When GENCO bids its actual cost, the cost of reactive power generation under LMP-based market is 6.84 times the cost

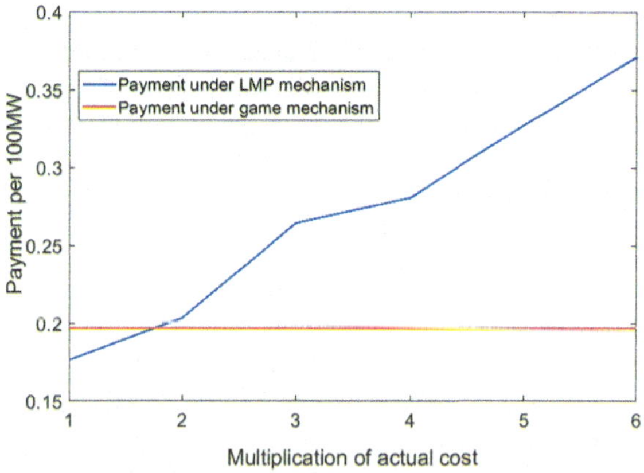

Fig. 6.4 Payment at bus 1 when cost function was magnified during bid

under game-based market. It is observed that when generation at bus 1 magnifies their actual cost of production by factors ranging from 1 to 6, cost of reactive power generation increases under LMP mechanism, whereas in game mechanism cost is much less.

The payment in game mechanism is compared with the LMP mechanism in Fig. 6.4. When GENCOs bid their actual cost, payment through LMP mechanism is 10.3% less than that through single-leader multi-follower game-based mechanism. On submitting bids higher than the actual cost, the payment increases in LMP mechanism. However, the generation company does not gain any additional profit on strategical bidding in single-leader multi-follower game model.

From Figs. 6.3 and 6.4, it can be observed that, while GENCOs exhibit a truth-telling behaviour, per unit price is much higher in game mechanism than that in LMP mechanism. Thus individual rationality is achieved. Under game mechanism, GENCO does not get any additional benefit by magnifying its cost function during bidding as it receives same profit. This demonstrates incentive compatibility of the proposed pricing mechanism.

6.6 Summary and Takeaways

In this chapter, the analysis of market mechanisms through existence of equilibrium, incentive compatibility and individual rationality of the single-leader multi-follower game-based market mechanisms for reactive power are studied in detail with illustrations in simple test systems. The theoretical results ensure that the proposed mechanisms provide an economically efficient pricing signal for reactive power market. Incentive compatibility of a market mechanism ensures that the model is capable of mitigating exercising of market power through financial withholding and capacity withholding. This is because truth-telling behaviour is induced through the pricing mechanisms among participants. Also, when individual rationality is preserved, participants earn a non-negative profit through the market mechanism by bidding their true type. Thus the market mechanisms encourage reactive power sources to participate in the three-stage market mechanism, provide required voltage support in the network and ensure safe and secure operation of the grid. Analysis of market mechanisms based on incentive compatibility and individual rationality is thus essential to understand the market efficiency of the pricing mechanism.

References

Fudenberg D, Jean T (1991) Tirole: game theory. MIT Press, pp 726–764
Kulkarni AA, Shanbhag UV (2014) A shared-constraint approach to multi-leader multi-follower games. Set-valued Variational Anal 22(4):691–720

Chapter 7
Deep Learning Analysis

Abstract This chapter presents the basics of deep learning required for analysis of bidding strategies in market mechanisms.

Keywords Deep reinforcement learning · Offline reinforcement learning · Long short-term memory network · Market bidding

This chapter presents the basics of deep learning required for analysis of bidding strategies in market mechanisms.

7.1 Basics of Deep Learning

Deep learning is way of classifying, clustering and predicting things network that has been trained on vast amounts of data. Deep learning is a set of learning methods attempting to model data with complex architectures. The elementary bricks of deep learning are the neural networks that are combined to form the deep neural networks. A major source of difficulty in many real-world artificial intelligence applications is that many of the factors of variation influence every single piece of data we are able to observe. Deep learning solves this central problem in representation learning by introducing representations that are expressed in terms of other, simpler representations. Deep learning allows the computer to build complex concepts out of simpler concepts. An example of a deep learning model is the feedforward deep network or multilayer perceptron (MLP). A mathematical function mapping some set of input values to output values is a multilayer perceptron. This function is formed by composing many simpler functions. Each application of a different mathematical function as providing a new representation of the input.

A multilayer perceptron (or neural network) as shown in Fig. 7.1 is composed by several hidden layers of neurons. In a multilayer perceptron, the output of a neuron of a layer becomes the input of a neuron of the next layer. Moreover, the output of a neuron can also be the input of a neuron of the same layer or of neuron of previous layers. The parameters of the architecture are the number of hidden layers and of

D. Jay and K. Shanti Swarup, *Market Operation for Reactive Power Ancillary Service*, Springer Tracts in Electrical and Electronics Engineering,
https://doi.org/10.1007/978-981-99-6952-4_7

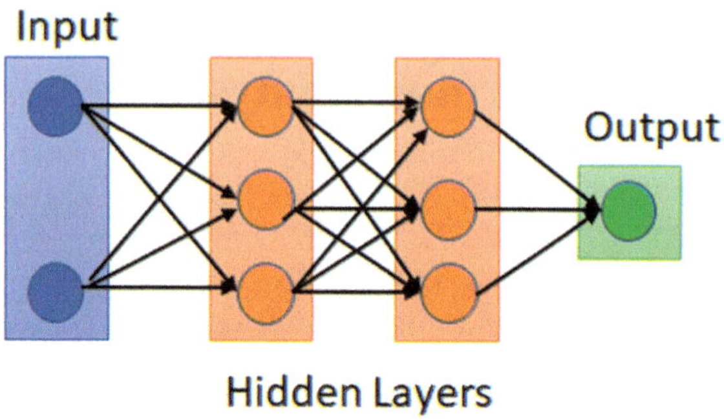

Fig. 7.1 Multilayer perceptron

neurons in each layer. The activation functions are also to be chosen by the user. On the last layer, called output layer, we may apply a different activation function as for the hidden layers depending on the type of problems we have at hand.

7.2 Classical Q-Learning

In order to solve reinforcement learning problem, the agent should learn the expected return for each state-action pair. Q-learning (Watkins and Dayan 1992) is a model-free, off-policy, temporal difference-based learning algorithm that uses Bellman Optimality Equation (7.1) to recursively update the Q-values until convergence is reached.

$$Q(s_t, a_t) = r_{t+1} + \gamma \max_{a'} Q(s_{t+1}, a') \tag{7.1}$$

Q-learning updates estimates based on learned estimates by bootstrapping. If we denote the learning rate as α, then the classical Q-learning update rule after k iterations is given by (7.2).

$$Q_{k+1}(s_t, a_t) = (1-\alpha)Q_k(s_t, a_t) + \alpha(r_{t+1} + \gamma \max_{a'} Q_k(s_{t+1}, a') \tag{7.2}$$

The term $\delta_t = r_{t+1} + \gamma \max_{a'} Q_k(s_{t+1}, a') - Q_k(s_t, a_t)$ is known as the TD-error, and it is a measure of the error in our current estimate of Q-values.

The exploration-exploitation dilemma is a much researched problem in reinforcement learning. If the agent determines the best action according to the current policy, then it may miss out on accurately computing the values of other states which could possibly give higher return. Hence, exploration is crucial for improving long-term return even though it might lower immediate rewards. ϵ-greedy method is used in this work for exploration.

In ϵ-greedy method, the greedy action based on the current policy is chosen most of the times and with a small probability ϵ 1, an action is randomly chosen from the action space. On approaching optimal action, ϵ is decayed over time.

To overcome the issues and limitation of classical tabular Q-learning batch mode Q-learning is used. In batch mode Q-learning, an approximation of Q-function over large state spaces is learnt with learning agent interacting with only the finite set of experience tuples (s_t, a_t, s_{t+1}, R_t). Fitted Q-iteration is a batch mode reinforcement learning algorithm that builds over the traditional value iteration and iteratively obtains a sequence of Q-function approximations. Basis functions like Fourier and radial basis functions (RBFs) are needed to map the states to the derived features space which will be used to approximate optimal Q-values. At each step, the learning agent has access to the experience buffer from which it samples a mini-batch as its training set. The algorithm is given below:

Algorithm 1: Fitted Q-Iteration

```
Initialize π_φ to a random policy
Generate samples (s_t, a_t, s_{t+1}, R_t)
while π_φ improves do
    y_t = r_t + Γmax_{a'} Q_φ(s_{t+1}, a')
    φ = arg min_φ Σ_t(Q_φ(s_t,a_t)-y_t)^2
end
```

7.3 Long Short-Term Memory Network

Recurrent neural networks (RNNs) are useful to handle sequential information. It is called recurrent because they perform the same task for each sequence, with output being dependent on previous states. A typical recurrent neural network is represented in Fig. 7.2.

x_t is the input vector at time t. s_t is the hidden state at time t. The hidden state is calculated based on the input vector, and the previous hidden state and output at time t are as

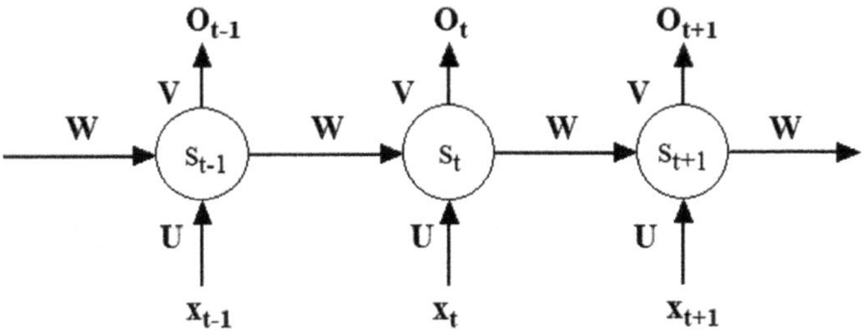

Fig. 7.2 Unrolled recurrent neural network

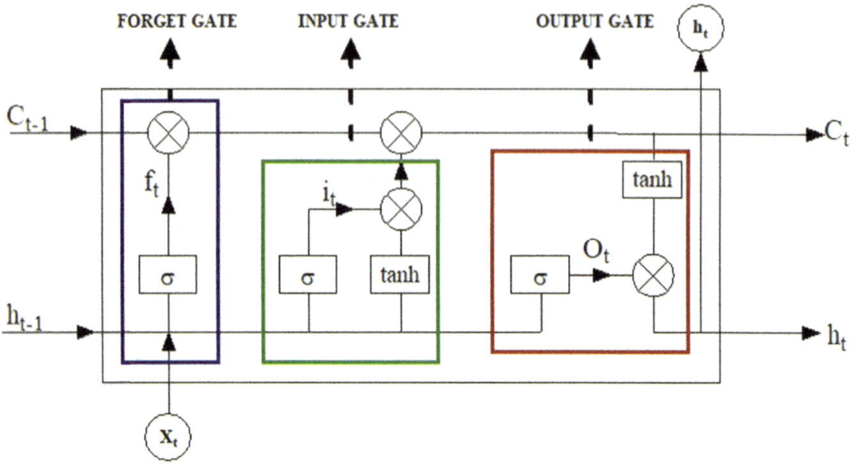

Fig. 7.3 Cell (repeating module) in an LSTM network

$$s_t = f(Ux_t + Ws_{t-1}) \tag{7.3}$$

$$O_t = f(Vs_t) \tag{7.4}$$

where f is the activate function like sigmoid function and ReLU.

However, the issue of vanishing gradient exists in RNNs. The dependency on previous state is propagated through the network as gradient. The lower the gradient, the harder the update of weights in the network. This vanishing gradient issue is solved through long short-term memory network (LSTM), as shown in Fig. 7.3. x_t is the input vector to the memory cell at time t and h_t is the value of memory cell at time t.

Input gate: The input gate determines which value from input should be used to modify the memory. i_t is the value of the input gate and which is given by

$$i_t = \sigma(W_i x_t + U_i h_{t-1} + b_i) \tag{7.5}$$

The candidate state of the memory cell at time t is formulated as

$$\tilde{C}_t = \tanh(W_c x_t + U_c h_{t-1} + b_i) \tag{7.6}$$

Forget gate: This gate discovers details that are to be discarded from the memory block. f_t is the value of forget gate, and C_t is the state of the memory cell at time t, which can be calculated by

$$f_t = \sigma(W_f x_t + U_f h_{t-1} + b_f); C_t = i_t \cdot \tilde{C}_t + f_t \cdot C_{t-1} \tag{7.7}$$

Output gate: The output is determined from the input and memory of the cell. O_t is the value of output gate and h_t is the value of memory cell at time t, as given by

$$O_t = \sigma(W_o x_t + U_o h_{t-1} + V_o C_t + b_o) \tag{7.8}$$

$$h_t = O_t \cdot \tanh(C_t) \tag{7.9}$$

7.4 Deep Learning-Based Analysis of Bidding Strategies

In real-time electricity markets, the objective of participants while bidding is to maximise their profit. The strategies for learning optimal bidding have been formulated through game theoretical approach and stochastic optimisation problem. In game theoretical approach, pay-offs of rivals are assumed to be known, which is unrealistic. In stochastic optimisation method, a suitable distribution function for rival's bids is assumed. The uncertainty in the network is handled by assuming the process to be Markov decision process with state transition probabilities known. Similar studies in reactive power market have not been reported so far because in reactive power markets, the network voltage operating conditions have an increased impact on reactive power markets than on active power markets. Also, contrary to active power markets, the bids of rivals are not directly related to fuel costs in reactive power markets. Hence assumption of a suitable probability distribution function is unrealistic. Thus strategies adopted in active power markets may not suit for learning optimal bids in reactive power market mechanisms. Hence, optimal bidding strategy is to be learnt from market observations and experience in imperfect oligopolistic competition-based markets. In this chapter, learning optimal bidding strategies from observation and experience in a three-stage reactive power market is presented. For learning optimal bidding strategies, a variant of neural fitted Q-iteration (NFQ-TP) with prioritised experience replay and target network is proposed. The total reactive power requirement is estimated by the learning agent through a long short-term memory (LSTM) network to suitably define the state space for the learning agent. The learning technique is tested on three-stage reactive power mechanism of IEEE 30-bus system under different scenarios. The simulation results ensure that the technique is suitable for learning optimal bidding strategies from own experiences and market observations.

7.4.1 Challenges in Learning Optimal Bidding Strategies

With the implementation of electricity markets based on competitive bidding, participants like generation companies, customers, etc. have been finding methods to determine optimal bidding strategies to utilise the varying load scenarios to earn profit. Several works related to optimal bidding strategies formulated the learning technique as an optimisation problem that was solved using heuristic methods (Wen and David 2001; Badri and Rashidinejad 2013). To consider the uncertainty of rival's behaviour while determining bidding strategies, minimisation of normal probability distribution function of the rival's bid is generally adopted. This minimisation problem has been solved using gravitational search algorithm (Singh and Fozdar 2019). Determining monthly bidding strategies was formulated as a bi-level problem in Jiang et al. (2019). Based on conditional value at risk, a stochastic bi-level optimisation was proposed (Rayati et al. 2019) for coordinated wind power and gas turbine units in the real-time market.

Day-ahead bidding problem can also be considered as a Markov decision process (MDP). In Wozabal and Rameseder (2020), a variant of the stochastic dual dynamic programming algorithm was used to solve as MDP. With advancements in artificial intelligence, reinforcement learning has found significant application in electricity market modelling (Nanduri and Das 2007; Rahimiyan and Mashhadi 2010). Recently, Deep Deterministic Policy Gradient with prioritised experience replay was proposed for complex energy markets (Ye et al. 2019). Asynchronous advantage actor-critic method was proposed in Cao et al. (2020) to determine optimal bidding strategies for wind farms in short-term electricity markets.

The aforementioned methodologies that determine the optimal bidding strategies are suitable for participants in energy markets. There has been no significant work related to formulation of optimal bidding strategy for participants in reactive power markets. This is mainly because of the complexity associated with reactive power market. These complexities arise with difficulty in determining the actual requirement of reactive power in the system. In the case of energy markets, the quantity to be produced by the participant can be directly determined from load forecasting techniques due to active power load pattern features. Whereas in reactive power markets, the requirement of reactive power in the system is determined not only by reactive power loading pattern but also other features like active power loading, network topology, system operating conditions like voltage, frequency, etc. and also the location of participant. Also estimating the cost function of rivals through probability distribution function or supply function methods is not easy in the case of reactive power markets. This is because the cost function of market participants is not directly related to fuel, contrary to energy markets. Thus, these issues lead to the requirement of formulating a new framework for learning optimal bidding strategies for participants in reactive power market.

Hence a learning technique that suits reactive power markets is given in Fig. 7.4.

The learning agent (participant) acts on the environment (reactive power market mechanism) by submitting their optimal bids. The feedback that it receives from the environment is the payment, which is the reward signal. Inputs that the learning agent requires to learn an optimal bidding strategy are market observations and experience. This defines the framework of the learning technique that is detailed in the following sections:

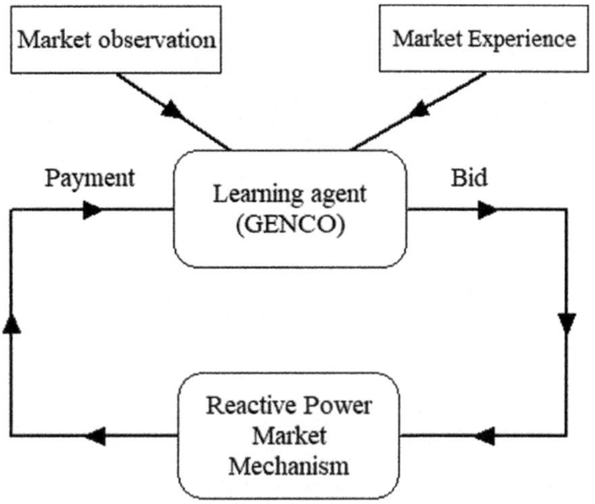

Fig. 7.4 Learning optimal bidding strategies in reactive power market

7.4.2 Reactive Power Market Environment for Learning

As described in earlier sections, learning of optimal bidding strategy in reactive power market becomes practical only when the learning process considers the following features for market environment

1. Higher-order Markov decision process.
2. Imperfect information setup.
3. Imperfect competition.

Higher-order MDP feature and imperfect information in the market are to be handled so that the market environment translates to a reinforcement learning setup. The feature of imperfect competition is discussed in detail in IEEE 30-bus system simulation studies.

Modelling of reactive power market environment to suit reinforcement learning setup is presented in this section. This section describes key definitions crucial to discussing methodologies presented in this work and formulates a Markov decision process (MDP) for the optimal bidding problem.

7.4.3 Handling Higher Order Markov Decision Process

This section describes key definitions crucial to the discussion of methodologies and formulates a Markov decision process (MDP) for the optimal bidding problem. In a reinforcement learning setting, an MDP is characterised by a set of states $\mathcal{S}$, a set of actions $\mathcal{A}$, transition probability function $p(s_{t+1}|s_t, a_t)$ satisfying Markov property that describes the probability of the learning agent being in state s_{t+1} on taking action a_t in state s_t and an immediate reward function $S \times A \to R$. At any time step, the goal of the learning agent is to choose an action a_t according to a policy $\Pi^*(s_t)$ that maximises the expected return $G_t = r_{t+1} + \Gamma r_{t+2} + \Gamma^2 r_{t+3} + \ldots$ where Γ is the discount factor that determines the trade-off between long-term and short-term rewards. This is done by following a policy that maximises the value of action value function $Q(s_t, a_t) = E[G_t|s_t, a_t, \Pi]$ for the given state s_t.

Translating the above to proposed market setting, the participant (e.g. GENCO) for which optimal bidding strategy is to be learnt from market observations constitutes the learning agent. Stages 2 and 3 of the three-stage market mechanism in Chap. ?? can be encapsulated into an optimisation model such that when any participant, say GENCO, submits its price bids to the model (stage 1), it receives the price signal and generation schedule for the next hour. This optimisation model is the environment in the reinforcement learning (RL) setting.

Under the assumption that reactive power requirement at any given time step remains almost the same irrespective of participants' behaviour, by fixing the bids of all participants to an arbitrary value and calculating the total quantity generated by all participants in the market at each time step, Figure 7.5 shows how total quantity generated by all reactive power producers (a measure of reactive power requirement) varies with time across an episode in IEEE 30-bus system. The dotted red lines are separated temporally by 24 h. A heavy correlation in loads at time step $t-48, t-24, t-2, t-1$ to the current load is observed. Therefore, it is possible to evaluate the price signals and generation schedule in a similar market situation from our experience and use them to predict the optimal behaviour in the current market scenario.

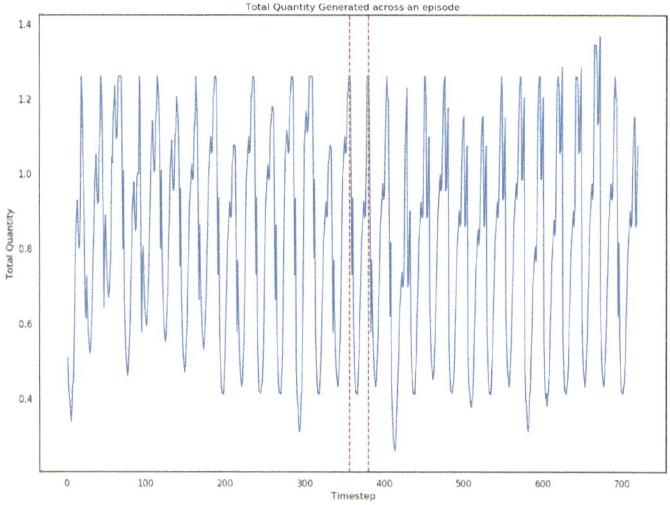

Fig. 7.5 Total quantity generated by all participants across an episode

Utilising signals from preceding time steps to describe the next state vector means that the probability of transitioning into the next state (time t) depends not only on the current state (time $t - 1$), but also on the sequence of states preceding it, specifically $t - 2, t - 24, t - 48$ in this case. This violates the first-order Markov assumption that *future is independent of the past given the present* which presents a major hurdle in formulating the problem as an MDP. However, such seemingly non-Markov cases such as ours when the next step depends on a bounded sequence of steps from past, i.e. $p(s_{t+1}|s_t, a_t, s_{t-1}, a_{t-1}, \ldots) = p(s_{t+1}|s_t, a_t, \ldots, s_{t-48}, a_{t-48})$, then such a higher order Markov process can be translated to a first-order Markov problem by augmenting the state vector and embedding the past (48 h) as an additional feature. Doing so makes the next step depend only on the current state and the action taken thereby making it possible to formulate an MDP for the problem.

7.4.4 Handling Imperfect Information in the Market

As mentioned in earlier sections, the reactive power requirement in the system depends on not only the reactive power loads in the network but also the network topology. This poses imperfect information setup in the learning environment. Load prediction techniques and probability-based assumptions will not be a direct input to determine the total quantity of reactive power generation required in the system. Thus, the reactive power requirement at each time step is estimated from the total quantity of reactive power generated in the network in previous time steps to see how the reactive power requirement at current time step correlates to the market experience we have gained so far.

For this, consider the time series data of total quantity generated by all GENCOs in IEEE 30-bus system from an earlier episode (month) as the raw training input. From the market analysis in Sect. 7.4.3, it is known that this data is periodic in nature which makes it suitable for using recurrent neural networks (RNNs), which is a class of neural networks that allow information to persist and hence make it easier to learn sequential data. An unrolled RNN can be thought of as multiple sequential copies of an artificial neural network. However, this poses an issue when it comes to long-term memory. RNNs consider only recent information while forgetting temporally separated events that could link to current output. This is where long short-term memory (LSTM) networks, a special class of recurrent neural networks, come in to learn long-term dependencies.

LSTMs accumulate an internal state that is constantly updated using the inputs, a hidden state that can be used for computing output, forget gate that determines what information to throw away from the cell state and together, the hidden state, input and cell state can be used to compute the next hidden state. Many-to-one LSTM models take a series of recent observations as input and generate a single output corresponding to prediction for next time step. Hence, the learning agent could use LSTMs to forecast load for the next hour as a sequence prediction.

The raw reactive power requirement (total quantity) signal is transformed into the training set for LSTM by splitting the input into sequence of length 24 corresponding to the reactive power requirement observations over the past one day. The target is defined as the total quantity for the next hour. Inputs from the training set are then fed into an LSTM network with n units followed by a dense layer. The network is trained by back-propagating the mean squared loss between predicted reactive power requirement and true reactive power requirement. The pretrained LSTM network can be used to estimate the total quantity required in the current state for state representation of the learning agent. The LSTM reactive power requirement prediction algorithm is outlined in Algorithm 2.

Algorithm 2: Reactive Power Requirement Prediction using LSTM

Let $S \leftarrow$ Total quantity time series data for a month
Initialize $X, y \leftarrow$ Split S into sequence of length 24 and reactive power requirement for next hour as corresponding output
Define model $M \leftarrow$ [LSTM(n), Dense]
for $iteration = 1, 2, \ldots$ **do**
  Sample a batch B from the training set Use M to generate the predicted output values $\hat{y}_i$ for $i = 1..B$ Train the network using Adam optimizer on $\sum_i (y_i - \hat{y}_i)^2$ for $i = 1..B$ and update the LSTM weights
end

7.4.5 *Features of Reinforcement Learning-Based Agent*

Having performed suitable translation of higher-order MDP to first-order MDP and handling imperfect information setup in the market environment, the features of the reinforcement learning-based agent are described as an MDP (S, A, P, R, Γ).

State Space: The state features are designed in such a way that the agent can use them to determine the consequences of its actions while also integrating the current market conditions like the load estimate.

1. **Previous bids**: This considers the past experience of the learning agent in the market to learn the optimal bidding strategy. We consider the ratio of bids sent to ISO to actual cost function of the learning agent i: $\langle a1_{i,t'}, a2_{i,t'} \rangle \forall t' \in \{t-48, t-24, t-2, t-1\}$ time stamps.
2. **Reward signals**: The rewards received at $\{t-48, t-24, t-2, t-1\}$ time steps act as feedback signal to the agent and will help the learner to judge the optimality of bids. A lower reward will indicate that the learner bid too high resulting in lesser payment. This is because the generation scheduled through stages 2 and 3 of the market mechanism was less due to high bid.
3. **Total Quantity Estimate**: We use the previous quantity signals to estimate the total generation at the current time step, which is correlated to the reactive power

requirement at that time step. We will later show how an LSTM could be utilised to predict this periodic quantity from time series data.

Action Space: The bids that we send to the optimisation module constitute the action for that state. For ease of analysis, instead of considering bids as action signals directly, we consider the action to be bid magnification defined as the ratio of bids sent to ISO to actual production cost. Thus for ith learner at time step t with actual operation cost $c1_i$ and lost opportunity cost $c2_i$, the action space is $(\mathrm{a}1_{i,t}, \mathrm{a}2_{i,t})$ defined as:

$$\langle a1_{i,t}, a2_{i,t} \rangle = \left\langle \frac{b1_{i,t}}{c1_i}, \frac{b2_{i,t}}{c2_i} \right\rangle, \tag{7.10}$$

where $b1_{i,t}$ and $b2_{i,t}$ represent the bid submitted to ISO as the claimed operation and lost opportunity cost, respectively. An advantage of using the ratios over actual bid values is that the actions can now remain same across all GENCOs leading to easier analysis. Because our action space is low dimensional, it is possible to discretise it at the required granularity and can be effectively considered as discrete space. In order to bound the action space, we restrict the bid magnification coefficients to $[1, 5]$ range and discretise the action space using step size of 0.5. This results in a total of 81 discrete actions: $(1, 1), (1, 1.5), \ldots, (2.5, 2.5), \ldots, (5, 4.5), (5, 5)$.

Reward Function: The reward is measured in terms of profit made by reactive power producer as compared to bidding its true cost. Let $c1_i$ and $c2_i$ represent the cost coefficients for agent i, qg be the quantity generated by the producer and bg_i be the base generation. Base generation of a generation company is the amount of reactive power required for the auxiliary services within the plant and also for shipment of base active power produced. Thus, we express the profit of GENCO as follows:

$$p_{i,t} = \text{price}_{i,t} \times qg_{i,t} - c1_i \times (qg_{i,t} - bg_i) - c2_i \times (qg_{i,t} - bg_i)^2 \tag{7.11}$$

The reward signal is computed as $r_{i,t} = p_{i,t} - p^b_{i,t}$, where $p^b_{i,t}$ is computed using price and generation returned from the environment upon sending the bids $(b1_{i,t}, b2_{i,t}) = (c1_i, c2_i)$ to the ISO optimisation model. By using a reward measure relative to baseline, the learner would have an incentive to learn actions that perform better than bidding their true costs and also be penalised by a negative reward when they receive payment less than baseline.

7.4.6 *Neural Fitted Q-Iteration*

Classical Q-learning algorithm suffers from the curse of dimensionality. The 13-dimensional state space defined for the learning agent is continuous and discretisation of it leads to exponential increase in state space size. Thus classical Q-learning technique is unsuitable for the learning agent in imperfect, oligopoly reactive power market. Hence deep reinforcement technique like neural fitted Q-iteration is required for learning optimal bidding strategies. Also, updating the Q-function based on only one point can lead to slower and noisy convergence.

Basis functions are needed to map the input states into the derived features space which will be used to approximate optimal Q-values. Any supervised learning technique, such as regression or SVM, can be used to fit the training set and generalise this information to any unseen state. At each step, the learning agent has access to the experience buffer from which it samples a mini-batch as its training set. Advantage of doing this is that the same experience tuple for learning multiple times while also breaking temporal correlations in sequentially generated data, thereby satisfying independently identically distributed (IID) assumption. The training set is of the form $(\langle s, a\rangle, Q)$ with the state-action pairs as input to the algorithm and the target being defined as the action value $Q(s, a)$ for the state-action pair.

Neural fitted Q-iteration (NFQ) (Riedmiller 2005) is a data efficient, deep learning-based algorithm belonging to the FQI family. It builds on the FQI algorithm by using the global generalisation effects of a neural network as a regressor to approximate the Q-values using offline data. The principle of NFQ is the same: using a single point to update the weights can create unintended changes in the weights for other state-action pairs whereas a batch-based approach would stabilise learning. NFQ uses resilient back-propagation (Rprop) as an optimiser to minimise oscillations in back-propagation. The advantage of using Rprop (Riedmiller and Braun 1993) is that it adapts the step size for each weight independently based on the sign of double derivative of loss function. Summarising, if Υ are the weights of the NFQ network, then corresponding to the input state-action pair $\langle s, a\rangle$, the output of the network would be the action value $Q(s, a; \Upsilon)$. Since NFQ has been shown to perform well in control problems with continuous state spaces, it is suitable for learning the Q-value function for the optimal strategy problem.

7.4.7 NFQ-TP-Based Learning Algorithm

In this section, variation of NFQ algorithm, i.e. NFQ with target network and prioritised experience replay *NFQ-TP* is proposed. The advantage is that the proposed algorithm leads to faster convergence for the bidding environment. The learning workflow for overall algorithm is described in Fig. 7.6.

In each learning iteration, the prioritised experience replay is replenished with new experiences by stepping in the environment in an ϵ-greedy manner by using the Q-value estimate from target network to find the maximising action. The weights of the local network are updated by sampling a mini-batch, using the local network to estimate Q-values for each experience tuple in the batch and performing back-propagation on mean squared TD loss with TD target computed using the target network. The weights of the target network are updated slowly based on the updates on the local network. This process is repeated until convergence is reached.

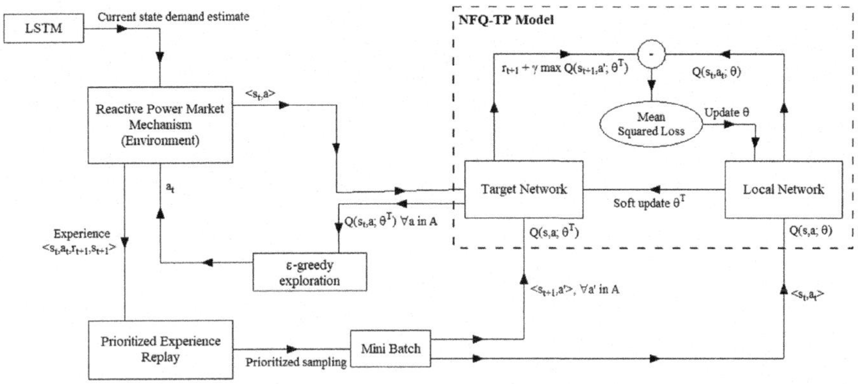

Fig. 7.6 Optimal bidding strategy training flow

7.5 Summary and Takeaways

The three-stage reactive power market model is designed to encourage GENCOs to bid closer to their true cost and prevent any single agent from gaining market power. The invisibility of rival bids and difficulty in estimating bid magnification from the price signals received by ISO present challenges in directly estimating the optimal bid value. Hence the learning environment proposed in this chapter takes features directly from the market observations such as bids and the corresponding rewards and learns using experience. Market behaviour and optimal bidding strategy are affected by the cost category of each generation company. While high and intermediate-cost GENCOs find it profitable to bid their actual cost in most of the scenarios, low-cost GENCOs make more profit than telling the truth. This is because low-cost GENCOs are free to bid higher in the common market so as to match their rivals actual cost. This suggests that separate reactive power markets for low-cost GENCOs like renewable energy sources would result in incentive compatibility, thereby leading to lower total cost for ISO.

References

Badri A, Rashidinejad M (2013) Security constrained optimal bidding strategy of Gencos in day ahead oligopolistic power markets: a Cournot-based model. Electr Eng 95(2):63–72

Cao D, Hu W, Xu X, Dragičević T, Huang Q, Liu Z, Chen Z, Blaabjerg F (2020) Bidding strategy for trading wind energy and purchasing reserve of wind power producer—a DRL based approach. Int J Electr Power Energy Syst 117:105648

Jiang Y, Hou J, Lin Z, Wen F, Li J, He C, Ji C, Lin Z, Ding Y, Yang L (2019) Optimal bidding strategy for a power producer under monthly pre-listing balancing mechanism in actual sequential energy dual-market in China. IEEE Access 7:70986–70998

Nanduri V, Das TK (2007) A reinforcement learning model to assess market power under auction-based energy pricing. IEEE Trans Power Syst 22(1):85–95

Rahimiyan M, Mashhadi HR (2010) An adaptive q-learning algorithm developed for agent-based computational modelling of electricity market. IEEE Trans Syst Man Cybern Part C (Appl Rev) 40(5):547–556

Rayati M, Goodarzi H, Ranjbar A (2019) Optimal bidding strategy of coordinated wind power and gas turbine units in real-time market using conditional value at risk. Int Trans Electr Energy Syst 29(1):e2645

Riedmiller M (2005) Neural fitted q iteration—first experiences with a data efficient neural reinforcement learning method. In: European conference on machine learning. Springer, Berlin, pp 317–328

Riedmiller M, Braun H (1993) A direct adaptive method for faster backpropagation learning: the RPROP algorithm. In: IEEE international conference on neural networks. IEEE, pp 586–591

Singh S, Fozdar M (2019) Optimal bidding strategy with the inclusion of wind power supplier in an emerging power market. IET Gener Transm Distrib 13(10):1914–1922

Watkins CJ, Dayan P (1992) Q-learning. Mach Learn 8(3–4):279–292

Wen F, David AK (2001) A genetic algorithm based method for bidding strategy coordination in energy and spinning reserve markets. Artif Intell Eng 15(1):71–79

Wozabal D, Rameseder G (2020) Optimal bidding of a virtual power plant on the Spanish day-ahead and intraday market for electricity. Eur J Oper Res 280(2):639–655

Ye Y, Qiu D, Sun M, Papadaskalopoulos D, Strbac G (2019) Deep reinforcement learning for strategic bidding in electricity markets. IEEE Trans Smart Grid 11(2):1343–1355

Chapter 8
Reactive Power Ancillary Service in Indian Grid Operation

Abstract This chapter discusses the aspects of reactive power management that are in practice in India and the feasibility of the proposed value-based market mechanism for reactive power ancillary service.

Keywords Ancillary services · Day-ahead market · Real-time market · Voltage profile

This chapter discusses the aspects of reactive power management that are in practice in India and the feasibility of the proposed value-based market mechanism for reactive power ancillary service.

8.1 Ancillary Services in India

In India, real-time operation of grid, grid operation studies, handling short-term open access market, etc., are handled by Power System Operation Corporation (POSOCO) at National Load Dispatch Centre (NLDC) for entire grid. There are five regional load dispatch centres under NLDC that control and regulate the real-time regional-grid activities. These grid operations are regulated by the Central Electricity Regulatory Commission and the operations follow the Indian Electricity Grid Code Regulations (IEGC).

Ancillary services have been defined in IEGC, 2010 Regulation 2(1) (b) that ancillary services are those services which are necessary to support the power system operation in maintaining power quality, reliability and security of the grid. The critical ancillary services identified by the regulation were active power support for load following, reactive power support and black start support.

The CERC (Unscheduled Interchange) Regulations, 2009 mandated NLDC to provide ancillary services. In the approach paper on ancillary services submitted by NLDC to CERC, three major categories of ancillary service were identified, namely

1. Load-generation balancing service.
2. Network control ancillary service.
3. System restart ancillary service.

D. Jay and K. Shanti Swarup, *Market Operation for Reactive Power Ancillary Service*, Springer Tracts in Electrical and Electronics Engineering,
https://doi.org/10.1007/978-981-99-6952-4_8

For load-generation balancing service, Reserve Regulation Ancillary Service (RRAS) was decided by CERC. With this the tertiary frequency control is achieved by utilising un-despatched, surplus capacity available in Inter-State Transmission System (ISTS). RRAS was launched by the NLDC in coordination with Regional LDCs on 18 April 2016. In December 2013, the Southern Grid was synchronised with the North-East-West Grid, thereby making One India One Frequency grid. This interconnection has facilitated economy interchange across regions, thereby replacing costlier power in deficit region with cheaper power from surplus region. RRAS has aided system operator to ensure secure and reliable grid operation by facilitating flexible operation in existing generation.

The approach paper also recommended the utilisation of hydrostations as synchronous condenser for providing reactive power support. A predetermined pricing formula has been in existence since 2010, to regulate the reactive power drawl/return during different voltage levels at the ISTS connection points. However, a market mechanism for reactive power support has not been envisaged.

8.1.1 Reactive Power Pricing in India

In India, according to Indian Electricity Grid Code Section 6.6, regional entities are bestowed with the responsibility of managing reactive power locally. Generation companies are not involved in the pricing methodology. The pricing formulation is as follows:

1. When voltage at metering point is below 97%, the regional entity pays for VAr drawl at Inter-State Transmission System (ISTS).
2. When voltage at metering point is below 97%, the regional entity gets paid for VAr return at (ISTS).
3. When voltage at metering point is above 103%, the regional entity pays for VAr return at ISTS.
4. When voltage at metering point is above 103%, the regional entity is paid for VAr drawl at ISTS.

A weekly statement of reactive energy charges giving details of payables/receivables with the pool and between beneficiaries is prepared by regional power committee. The rate for charge/payment for reactive energy is governed as per clause 6.6 (2) of Indian Electricity Grid Code (CERC) Regulations, 2012 which shall be 10 paise/KVArh and shall be escalated at 0.5 paise/KVArh per year thereafter. Presently, the rate is 15 Paise/KVArh for Year 2020–2021.

8.2 Drawbacks in Existing Reactive Power Pricing

The voltage profile in several specific buses in Southern Region (SR) grid has been observed from the data available at SRLDC website. Voltage deviation index (VDI) is calculated as the ratio of number of hours at high voltage to 24 h. A typical voltage profile at Alamathy bus, Tamil Nadu, in the month of December for two financial years is tabulated in Table 8.1. It can be observed from Table 8.1 that during high voltage there is MVAR absorption at the station. This will result in amount receivable at Tamil Nadu from pool. However, the voltage profile at the bus was not within the permissible limit. Similarly, there have been many instances where MVAr injection has been observed during high-voltage conditions and the regional entity pays the pool.

pagination

The buses that were identified for this observation are Tirunelveli, Karaikudi, Pugalur, Alamathy and Kalivanthapattu. MVAr injection is observed from lines between Vallore TPS-Alamathy and Vallore TPS-Kalivanthapattu. From MVAR data at Alamathy, Vallore-Alamathy lines have MVAr absorption profile. However, the bus voltage was observed to be higher than the IEGC band almost 18–24 h a day. From MVAR data at Kalivanthapattu Vallore TPS—Kalivanthapattu lines have MVAr injection profile. Thus Tamil Nadu was payable to the pool for MVAR injection during high-voltage profile.

Such situation has arisen because the reactive power injection/absorption at Vallore TPS is not well coordinated, resulting in Tamil Nadu being payable to the pool for the MVAr injection during high-voltage profile of the grid. These observations indicate that the existing pricing mechanism on regional entities is not effective in managing reactive power in the grid.

8.3 Feasibility and Benefits of Value-Based Reactive Power Market

The value-based reactive power market mechanism will be suitable for effective reactive power management in the network. Feasibility analysis of the proposed value-based reactive power market can be carried out region-wise on the ISTS system. Initial analysis can be based on perfect competition environment which shall enable in identifying critical locations in the network that are reactive power deficits.

The feasibility analysis may include practical system constraints like reactive power reserve bounds, real-time active power market settlement, etc., in the value-based reactive power market. Also, the actual implementation of reactive power market mechanism involves policy and regulation changes. Such changes will include the time-frame for implementation. The proposed mechanism is found to be suitable for day-ahead and real-time market operations.

Table 8.1 Voltage deviation at Alamathy, Tamil Nadu

Date	FY 2018–2019		FY 2019–2020	
	MVAr (absorption)	VDI	MVAr (absorption)	VDI
Dec 1	2007.95	0.83	2338.91	1.00
Dec 2	2317.84	1.00	1400.03	1.00
Dec 3	1198.46	0.41	1455.27	1.00
Dec 4	1258.92	0.46	1299.60	1.00
Dec 5	1207.32	0.53	1447.28	1.00
Dec 6	1066.13	0.50	1648.72	1.00
Dec 7	1101.13	0.52	1791.26	1.00
Dec 8	1111.24	0.53	3052.36	1.00
Dec 9	1901.08	0.98	1805.85	0.94
Dec 10	1179.60	0.47	1666.16	0.85
Dec 11	888.75	0.32	1670.52	0.98
Dec 12	942.55	0.36	1701.86	0.72
Dec 13	994.94	0.42	1680.68	0.66
Dec 14	862.49	0.38	2963.66	1.00
Dec 15	1156.36	0.62	3505.45	1.00
Dec 16	2628.41	0.92	1980.35	0.93
Dec 17	1350.51	0.63	1698.19	0.94
Dec 18	1063.97	0.47	1483.64	0.53
Dec 19	1187.66	0.44	1381.06	0.81
Dec 20	1275.62	0.55	1261.09	0.49
Dec 21	1466.93	0.54	1500.38	0.93
Dec 22	1624.65	0.51	2912.72	1.00
Dec 23	3472.78	1.00	2074.89	0.83
Dec 24	1740.39	0.53	1976.74	0.79
Dec 25	1682.10	0.66	2432.01	1.00
Dec 26	1510.56	0.48	1862.52	0.89
Dec 27	1298.92	0.44	1697.45	0.78
Dec 28	1381.81	0.54	1408.73	0.69
Dec 29	1333.12	0.49	2299.64	1.00
Dec 30	2125.79	1.00	1670.54	0.98
Dec 31	1437.10	0.53	1395.63	0.76

For day-ahead market operations, the proposed value-based market mechanism may include system uncertainty consideration. This can be handled using techniques like Information Gap Decision Theory (IGDT) that will suit optimisation problems. Also, the influence of Value of Lost Load (VoLL) in reactive power market and the percentage of impact of VoLL in the reactive power market may be required to be studied region-wise. Region-wise analysis of these features is required as India is a diverse country; the nature of load also varies with geographical and climatic diversity.

In real-time market operation, the ex-ante and ex-post market settlement features are to be well defined upon implementation. This will also require extensive case studies on each region to arrive at an efficient regulatory policy. Also, unforeseen contingencies like outage of market participant and its impact on system performance will require extensive analysis. Based on such feasibility studies, penalties/compensation can be decided at regulatory level to handle such situations by the system operator.

Deciding suitable constraints for reactive power reserve requirement in the network will also play a significant role in the implementation of value-based reactive power reserve. In the studies performed on IEEE test systems such reactive power reserve constraints were not considered. This was because the reactive power capacity in such test cases is sufficient to handle the network contingencies. The simulation studies could thus only analyse whether the reactive power reserves are not traded by the market participants to earn profit in the network during severe outage cases. However, in actual implementation of value-based market mechanism, constraints on reactive power reserve during contingencies and its impact on price volatility may also need a detailed analysis.

Though the actual implementation of value-based reactive power market mechanism requires further extensive studies, which are specific to each region in the national grid, several benefits are expected with system studies on the proposed market mechanism. These benefits are as follows:

1. Identification of critical buses in the region which are reactive power deficits. This can be identified through locational marginal price analysis in value-based reactive power market under perfect competition.
2. Optimal reactive power schedule under normal operation can be achieved through the proposed reactive power value function-based AC-OPF.
3. With the increased penetration of private firms in power generation sector, incentive compatible value-based reactive power market mechanisms shall be suitable to encourage them to participate in the market and at the same time curtail exercise of market power.
4. Individual rationality of market mechanism encourages participants to invest in providing more dynamic reactive power sources in the network, which are suitable for real-time market.
5. Solar/wind power plants, which are potential reactive power sources, are encouraged to participate in the market. At present, these sources are not being utilised due to Power Purchase Agreement commitments and penalties. Individual rationality feature of the value-based market mechanism ensures that such potential sources are provided optimal price signals to encourage their participation.

The immediate benefit of the proposed value-based market formulation under perfect competition in Indian grid is that critical locations in the network can be identified. This helps the grid operators to effectively manage the existing resources by applying suitable load management, demand forecasting, scheduled outage management, etc. It can also help in providing operational feedback to the system planners for investment in the critical locations to cater the reactive power requirement in the system. This will help in improving the system operating conditions in general. However, an investment in reactive power market can be encouraged with market mechanisms. Hence implementation of market mechanism for reactive power ancillary service is also essential in Indian grid from operational as well as network expansion planning perspective.

8.4 Summary and Takeaways

This chapter provided the impact of proposed value-based reactive power market mechanism in Indian grid. It is understood that lack of efficient pricing mechanism for reactive power has resulted in several operational issues like localised high voltage and forced outage of critical lines. An efficient nodal pricing mechanism will facilitate grid operators in identifying critical locations in the network, which are reactive power deficits. This can help in providing operational feedback to the Central Transmission Utility (CTU-planning), thereby executing suitable network expansion planning. Effective utilisation of private reactive power sources, dynamic VArs and renewable energy plants can be achieved through suitable pricing mechanisms like value-based reactive power market mechanism, which is incentive compatible and preserves individual rationality. A non-negative profit encourages investment in the network which will improve the system operating conditions in the long run.

Appendix A
GAMS Code

A.1 DC-OPF for IEEE 9-Bus System

```
Set
i 'network buses' / 1*9 /
slack(i) / 1 /

   Scalar
Sbase / 100 /;

   Alias (i, j);

   ________________________________

Table GenD(i, *) 'generating units characteristics'
pmax pmin Qmax Qmin a b c
1 250 10 300 −300 0.11 5 150
2 300 10 300 −300 0.085 1.2 600
3 260 10 300 −300 0.1225 1 335;

   Table BD(i, *) 'demands of each bus in MW'
Pd Qd
5 125 0
6 90 0
8 100 0;
****************************************************

   Table LN(i, j, *) 'network technical characteristics'
x b limit
1 .4 0.0576 0 250
4 .5 0.068 0.158 250
4 .6 0.092 0.158 250
```


D. Jay and K. Shanti Swarup, *Market Operation for Reactive Power Ancillary Service*, Springer Tracts in Electrical and Electronics Engineering,
https://doi.org/10.1007/978-981-99-6952-4

```
5 .7 0.161 0.306 250
2 .7 0.0625 0 250
7 .8 0.0576 0.00674 300
8 .9 0.1008 0.209 150
6 .9 0.1738 0.358 150
3 .9 0.0586 0 300;

    ****************************************************

    *LN(i, j, 'b') = 0;
LN(i, j, 'x')(LN(i, j, 'x') = 0) = LN(j, i, 'x');
LN(i, j, 'r')(LN(i, j,'r') = 0) = LN(j, i,'r');
LN(i, j,'b')(LN(i, j, 'b') = 0) = LN(j, i, 'b');
LN(i, j, 'Limit')(LN(i, j, 'Limit') = 0) = LN(j, i, 'Limit');
LN(i, j,'bij')LN(i, j,'Limit') = 1/LN(i, j, 'x');
LN(i, j, 'z')LN(i, j, 'Limit') = sqrt(sqr(LN(i, j, 'x')) + sqr(LN(i, j, 'r')));
LN(j, i,'z')(LN(i, j, 'z') = 0) = LN(i, j, 'z');
LN(i, j, 'th')(LN(i, j, 'Limit') and LN(i, j, 'x') and LN(i, j, 'r')) = arctan(LN(i, j,
'x')/(LN(i, j, 'r')));
LN(i, j, 'th')(LN(i, j, 'Limit') and LN(i, j, 'x') and LN(i, j, 'r') = 0) = pi/2;
LN(i, j, 'th')(LN(i, j, 'Limit') and LN(i, j, 'r') and LN(i, j, 'x') = 0) = 0;
LN(j, i, 'th')LN(i, j, 'Limit') = LN(i, j, 'th');
Parameter cx(i, j);
cx(i, j)(LN(i, j,'limit') and LN(j, i,'limit')) = 1;
cx(i, j)(cx(j, i)) = 1;
```

```
    Variable OF, Pij(i, j), Qij(i, j), Pg(i), Qg(i), Va(i), V(i);
Equation eq1, eq2, eq3, eq4, eq5, eq6, eq7;

    eq1(i, j)cx(i, j)..
Pij(i, j) = e = LN(i, j, 'bij')*(Va(i) - Va(j));

    eq2(i)..
Pg(i)GenD(i, 'Pmax') - BD(i, 'pd')/Sbase = e = sum(jcx(j, i), Pij(i, j));

    eq3..
OF = e = sum(i, Pg(i)*GenD(i, 'b')*Sbase + Pg(i)*Pg(i)*GenD(i, 'a')*Sbase*Sbase
+ GenD(i, 'c'));

    Model loadflow / eq1, eq2, eq3/;
```

```
Pg.lo(i) = GenD(i, 'Pmin')/Sbase;
Pg.up(i) = GenD(i, 'Pmax')/Sbase;

Va.up(i) = pi/2;
Va.lo(i) =−pi/2;
Va.l(i) = 0;
Va.fx(slack) = 0;

Pij.up(i, j)((cx(i, j))) = 1*LN(i, j, 'Limit')/Sbase;
Pij.lo(i, j)((cx(i, j))) = −1*LN(i, j, 'Limit')/Sbase;

V.lo(i) = 0.9;
V.up(i) = 1.1;
V.l(i) = 1;

solve loadflow minimizing OF using nlp;
```

A.2 AC-OPF for PJM 5-Bus System

```
Set
i 'network buses' / 1*5 /
slack(i) / 1 /

Scalar
Sbase / 100 /;

Alias (i, j);

* ____________________________________________
Table GenD(i, *) 'generating units characteristics'
pmax pmin Qmax Qmin b
1 210 0 157.5 −157.5 3
3 520 0 390 −390 5
4 200 0 150 −150 2
5 600 0 450 −450 1;

Table BD(i, *) 'demands of each bus in MW'
Pd Qd
1 0 0
2 300 98.61
3 300 98.61
4 400 131.47
5 0 0 ;
```

```
*******************************************************

   Table LN(i, j, *) 'network technical characteristics'
r x b limit
1 .2 0.00281 0.0281 0.00712 400
1 .4 0.00304 0.0304 0.00658 400
1 .5 0.00064 0.0064 0.03126 400
2 .3 0.00108 0.0108 0.01852 400
3 .4 0.00297 0.0297 0.00674 400
4 .5 0.00297 0.0297 0.00674 240;

   *******************************************************

   *LN(i, j, 'b') = 0;
LN(i, j, 'x')(LN(i, j, 'x') = 0) = LN(j, i, 'x');
LN(i, j, 'r')(LN(i, j, 'r') = 0) = LN(j, i, 'r');
LN(i,j,'b')(LN(i, j, 'b') = 0) = LN(j, i, 'b');
LN(i, j, 'Limit')(LN(i, j, 'Limit') = 0) = LN(j, i, 'Limit');
LN(i,j,'bij')LN(i, j, 'Limit') = 1/LN(i, j, 'x');
LN(i, j, 'z')LN(i, j, 'Limit') = sqrt(sqr(LN(i, j, 'x')) + sqr(LN(i, j, 'r')));
LN(j, i, 'z')(LN(i, j, 'z') = 0) = LN(i, j, 'z');
LN(i, j, 'th')(LN(i, j, 'Limit') and LN(i, j, 'x') and LN(i, j, 'r')) = arctan(LN(i, j,
'x')/(LN(i, j, 'r')));
LN(i, j, 'th')(LN(i, j, 'Limit') and LN(i, j, 'x') and LN(i, j, 'r') = 0) = pi/2;
LN(i, j, 'th')(LN(i, j, 'Limit') and LN(i, j, 'r') and LN(i, j, 'x') = 0) = 0;
LN(j, i, 'th')LN(i, j, 'Limit') = LN(i, j, 'th');

   Parameter cx(i, j);
cx(i, j)(LN(i, j, 'limit') and LN(j, i, 'limit')) = 1;
cx(i, j)(cx(j, i)) = 1;
```

```
   Variable OF, Pij(i,j), Qij(i,j), Pg(i), Qg(i), Va(i), V(i);
Equation eq1, eq2, eq3, eq4, eq5, eq6, eq7;

   eq1(i, j)cx(i, j)..
Pij(i, j) = e = (V(i)*V(i)*cos(LN(j, i,'th')) – V(i)*V(j)*cos(Va(i) – Va(j) + LN(j, i,
'th')))/LN(j, i, 'z');

   eq2(i, j)cx(i, j)..
Qij(i, j) = e = (V(i)*V(i)*sin(LN(j, i, 'th')) – V(i)*V(j)*sin(Va(i) – Va(j) + LN(j, i,
'th'))) /LN(j, i, 'z') – LN(j, i, 'b')*V(i)*V(i)/2;
```

```
    eq3(i)..
Pg(i)GenD(i, 'Pmax') – BD(i, 'pd')/Sbase = e = sum(jcx(j, i), Pij(i, j));

    eq4(i)..
Qg(i)GenD(i, 'Qmax') – BD(i, 'qd')/Sbase = e = sum(jcx(j, i), Qij(i, j));

    eq5..
OF =e= sum((i), Qg(i)*GenD(i, 'b')*SbaseGenD(i, 'Pmax'));

    Model loadflow / eq1, eq2, eq3, eq4, eq5 /;

    Pg.lo(i) = GenD(i, 'Pmin')/Sbase;
Pg.up(i) = GenD(i, 'Pmax')/Sbase;
Qg.lo(i) = GenD(i, 'Qmin')/Sbase;
Qg.up(i) = GenD(i, 'Qmax')/Sbase;

    Va.up(i) = pi/2;
Va.lo(i) =−pi/2;
Va.l(i) = 0;
Va.fx(slack) = 0;

    Pij.up(i, j)((cx(i, j))) = 1*LN(i, j, 'Limit')/Sbase;
Pij.lo(i, j)((cx(i, j))) = −1*LN(i, j, 'Limit')/Sbase;
Qij.up(i, j)((cx(i, j))) = 1*LN(i, j, 'Limit')/Sbase;
Qij.lo(i, j)((cx(i, j))) = −1*LN(i, j,'Limit')/Sbase;

    V.lo(i) = 0.9;
V.up(i) = 1.1;
V.l(i) = 1;

    solve loadflow minimizing OF using nlp;
```

A.3 Multi-period Value-Based Reactive Power Market for IEEE 24-Bus System

```
Set
i 'network buses' / 1*24 /
slack(i) / 13 /
k / 1, 2, 7, 13, 14*16, 18, 21*23 /
t / t1*t24 /;
Scalar
```

```
Sbase / 100 /;

   Alias (i, j);
   ******************************************************
Table GenD(i, *) 'generating units characteristics'
pmax pmin b Qmax Qmin Vg a0 a1 a2 a c QA QB R R1 lnk
1 152 30.4 13.32 192 −50 1.035 0.046 0.86 0.46 3 110 150 192 7.30 0.137 3
2 152 30.4 13.32 192 −50 1.035 0.039 0.68 0.39 4.05 100 150 192 6.84 0.146 3
7 350 75 20.70 300 0 1.025 0.043 0.75 0.43 4.05 180 250 300 12.37 0.081 1
13 591 206.85 20.90 591 0 1.02 0.050 0.60 0.50 0.99 250 500 591 12.58 0.079 3
14 0 0 0 200 −50 1.02 0.075 0.75 0.90 0 0 0 0 13.63 0.073 2
15 215 66.25 21.00 215 −100 1.014 0.065 0.73 0.38 3.99 98 180 215 13.47 0.074 4
16 155 54.25 10.00 155 −50 1.017 0.050 0.50 0.30 3.88 100 100 155 12.41 0.081 4
18 400 100 5.47 400 −50 1.05 0.090 0.85 0.48 4.5 150 350 400 13.94 0.072 3
21 400 100 5.47 400 −50 1.05 0.080 0.75 0.41 4.5 250 350 400 14.00 0.071 5
22 300 0 5.97 300 −60 1.05 0.055 0.48 0.38 3.9 190 250 300 14.61 0.068 2
23 360 248.5 10.50 310 −125 1.05 0.095 0.89 0.55 3.9 190 280 310 8.57 0.117 4;

   ******************************************************
Table GenQ(i, *) 'generating units characteristics'
R R1
1 2.12 0.473
2 1.89 0.527
7 11.12 0.089
13 9.68 0.103
14 10.08 0.099
15 9.39 0.106
16 8.52 0.112
18 9.65 0.104
21 9.72 0.103
22 10.12 0.098
23 5.58 0.179;

   ******************************************************

   Table BD(i, *) 'demands of each bus in MW'
Pd Qd Vref
1 110 20 1.035
2 97 22 1.035
3 180 37 1.000
4 74 15 1.000
5 71 14 1.000
6 136 28 1.000
7 125 25 1.025
8 171 35 1.000
```

```
9 175 36 1.000
10 195 40 1.000
11 0 0 1.000
12 0 0 1.000
13 265 54 1.020
14 194 39 1.020
15 317 64 1.014
16 100 20 1.017
17 0 0 1.000
18 333 68 1.050
19 181 37 1.000
20 128 26 1.000
21 0 0 1.050
22 0 0 1.050
23 0 0 1.050
24 0 0 1.000 ;
    ****************************************************

    Table LN(i, j, *) 'network technical characteristics'
r x b limit
1 .2 0.0026 0.0139 0.4611 175
1 .3 0.0546 0.2112 0.0572 175
1 .5 0.0218 0.0845 0.0229 175
2 .4 0.0328 0.1267 0.0343 175
2 .6 0.0497 0.192 0.052 175
3 .9 0.0308 0.119 0.0322 175
3 .24 0.0023 0.0839 0 400
4 .9 0.0268 0.1037 0.0281 175
5 .10 0.0228 0.0883 0.0239 175
6 .10 0.0139 0.0605 2.459 175
7 .8 0.0159 0.0614 0.0166 175
8 .9 0.0427 0.1651 0.0447 175
8 .10 0.0427 0.1651 0.0447 175
9 .11 0.0023 0.0839 0 400
9 .12 0.0023 0.0839 0 400
10.11 0.0023 0.0839 0 400
10.12 0.0023 0.0839 0 400
11.13 0.0061 0.0476 0.0999 500
11.14 0.0054 0.0418 0.0879 500
12.13 0.0061 0.0476 0.0999 500
12.23 0.0124 0.0966 0.203 500
13.23 0.0111 0.0865 0.1818 500
14.16 0.005 0.0389 0.0818 500
15.16 0.0022 0.0173 0.0364 500
15.21 0.00315 0.0245 0.206 1000
```

```
15.24 0.0067 0.0519 0.1091 500
16.17 0.0033 0.0259 0.0545 500
16.19 0.003 0.0231 0.0485 500
17.18 0.0018 0.0144 0.0303 500
17.22 0.0135 0.1053 0.2212 500
18.21 0.00165 0.01295 0.109 1000
19.20 0.00255 0.0198 0.1666 1000
20.23 0.0014 0.0108 0.091 1000
21.22 0.0087 0.0678 0.1424 500 ;
* ______________________________
Table WD(t, *)
w d
t1 0.0786666666666667 0.684511335492475
t2 0.0866666666666667 0.644122690036197
t3 0.117333333333333 0.6130691560297
t4 0.258666666666667 0.599733282530006
t5 0.361333333333333 0.588874071251667
t6 0.566666666666667 0.5980186702229
t7 0.650666666666667 0.626786054486569
t8 0.566666666666667 0.651743189178891
t9 0.484 0.706039245570585
t10 0.548 0.787007048961707
t11 0.757333333333333 0.839016955610593
t12 0.710666666666667 0.852733854067441
t13 0.870666666666667 0.870642027052772
t14 0.932 0.834254143646409
t15 0.966666666666667 0.816536483139646
t16 1 0.819394170318156
t17 0.869333333333333 0.874071251666984
t18 0.665333333333333 1
t19 0.656 0.983615926843208
t20 0.561333333333333 0.936368832158506
t21 0.565333333333333 0.887597637645266
t22 0.556 0.809297008954087
t23 0.724 0.74585635359116
t24 0.84 0.733473042484283;
* ______________________________

  LN(i, j, 'b') = 0;

  LN(i, j, 'x')(LN(i, j, 'x') = 0) = LN(j, i, 'x');
LN(i, j, 'r')(LN(i, j, 'r') = 0) = LN(j, i, 'r');
LN(i, j, 'b')(LN(i, j, 'b') = 0) = LN(j, i, 'b');
LN(i, j, 'Limit')(LN(i, j, 'Limit') = 0) = LN(j, i,'Limit');
LN(i, j, 'bij')LN(i, j, 'Limit') = 1/LN(i, j, 'x');
```

```
LN(i, j, 'z')LN(i, j, 'Limit') = sqrt(sqr(LN(i, j, 'x')) + sqr(LN(i, j, 'r')));
LN(j, i, 'z')(LN(i, j, 'z') = 0) = LN(i, j, 'z');
LN(i, j, 'th') (LN(i, j, 'Limit') and LN(i, j, 'x') and LN(i, j, 'r')) = arctan(LN(i, j,
'x')/(LN(i, j, 'r')));
LN(i,j,'th')(LN(i, j, 'Limit') and LN(i, j, 'x') and LN(i, j, 'r') = 0) = pi/2;
LN(i, j, 'th')(LN(i, j, 'Limit') and LN(i, j, 'r') and LN(i, j, 'x') = 0) = 0;
LN(j, i, 'th')LN(i, j, 'Limit') = LN(i, j, 'th');

   Parameter cx(i, j);
cx(i, j)(LN(i, j, 'limit') and LN(j, i, 'limit')) = 1;
cx(i, j)(cx(j, i)) = 1;

   Binary Variable A,B;
Variable OF, Pij(i, j, t), P(i, t), P1, P2, Qij(i, j, t), Qu(i), Pg(i, t), Qg(i, t),TL(i, j),
Va(i, t), V(i, t), Qm1(t), Qmm1(i, t), Qmex1, Qgl, Qmm2(i, t),Qrr1(t), Qm2, Qgl2,
Qrr2, Qmw(i, t), Qres(i, t), Qgl1(i, t), Qd1, Qg2, Qg1, Qd2, Qmm(i, t), Lij, Sij, Qm,
Ql1, Ql2, Qmex, Ql, Qa, QB, Qmm, Qlg, Qrr, dual(i), Qlg1(t), Qlg2;
Equation eq1, eq2, eq3, eq4, eq5, eq6, eq7, eq8, eq9, eq10, eq11, eq12, eq13;

   eq1..
OF = e = sum(t, ((Ql1(t) - Qm1(t))) - (Qrr1(t)) + Qlg1(t));
   eq2(i, j, t)cx(i, j)..
Pij(i, j, t) = e = (V(i, t)*V(i, t)*cos(LN(j, i, 'th')) - V(i, t)*V(j, t)*cos(Va(i, t) -
Va(j, t) + LN(j, i, 'th'))) /LN(j, i, 'z');

   eq3(i, j, t)cx(i, j)..
Qij(i, j, t) = e = (V(i, t)*V(i, t)*sin(LN(j, i, 'th')) - V(i, t)*V(j, t)*sin(Va(i, t) - Va(j,
t) + LN(j, i, 'th'))) /LN(j, i, 'z') - LN(j, i, 'b')*V(i, t)*V(i, t)/2;
eq4(i, t)..
Pg(i, t)GenD(i, 'Pmax') - WD(t, 'd')*BD(i, 'pd')/(Sbase) = e = sum(jcx(j, i),
Pij(i, j, t));
   eq5(i, t)..
Qg(i, t)GenD(i, 'Qmax') - WD(t, 'd')*BD(i, 'Qd')/(Sbase) = e = sum(jcx(j, i),
Qij(i, j, t));

   eq6(i, t)..
Qmw(i, t) = e = sum((j), (LN(i, j, 'x')*Pij(i, j, t)*Pij(i, j, t))/(V(j, t)*V(j, t)));

   eq7(i, t)..
Qg(i, t)GenD(i, 'Qmax') = e = (GenD(i, 'Qmax'))/Sbase - (Qres(i, t));

   eq8(t)..
Ql1(t) = e = sum(i, Qg(i, t)*GenD(i, 'a1')) ;
```

```
eq9(t)..
Qm1(t) = e = sum(i, Qmm(i, t)*GenD(i, 'a1'));

eq10(t)..
Qrr1(t) = e = sum(i, Qres(i, t)*Qres(i, t)*GenD(i, 'a2'));

eq11(t)..
Qlg1(t) = e = sum(i, Qmm(i, t)*(GenD(i, 'a1') + GenD(i, 'a2')*Qmm(i, t))*GenQ(i,
'R1'));
eq12(i, t)..
abs(V(i, t) − 1)/1 = l = 0.04;
eq13(i, t)..
((1) − (V(i, t)*V(i, t)) + 1)*sum(j, (LN(i, j, 'bij'))) = g = Qmm(i, t);
Model loadflow /eq1, eq2, eq3, eq4, eq5, eq6, eq7, eq8, eq9, eq10, eq11, eq12,
eq13/;

Pg.lo(i, t) = GenD(i,'Pmin')/Sbase;
Pg.up(i, t) = GenD(i,'Pmax')/Sbase;
Qg.lo(i, t) = GenD(i,'Qmin')/Sbase;
Qg.up(i, t) = GenD(i,'Qmax')/Sbase;

Va.up(i, t) = pi/2;
Va.lo(i, t) =−pi/2;
Va.l(i, t) = 0;
Va.fx(slack, t) = 0;
Pij.up(i, j, t)((cx(i, j))) = 1*LN(i, j, 'Limit')/Sbase;
Pij.lo(i, j, t)((cx(i, j))) = −1*LN(i, j, 'Limit')/Sbase;
Qij.up(i, j, t)((cx(i, j))) = 1*LN(i, j, 'Limit')/Sbase;
Qij.lo(i, j, t)((cx(i, j))) = −1*LN(i, j, 'Limit')/Sbase;
V.lo(i, t) = 0.95;
V.up(i, t) = 1.05;
V.l(i, t) = 1;
solve loadflow min OF using dnlp;
```

A.4 Single-Leader Follower Game Model in PJM 5-Bus

```
Set
i 'network buses' / 1*5/
slack(i) / 5 / ;
Scalar
Sbase / 100 /;
Alias (i, j);
```

```
Table GenD(i, *) 'generating units characteristics'
pmax pmin Qmax Qmin Vg a0 a1 a2 a c QA QB R R1 lnk Qbb
1 210 0 157.5 −157.5 1.0 0.046 0.86 0.46 3 110 115.2 192 7.30 0.047 3 15.0
3 520 0 390 −390 1.0 0.039 0.68 0.39 4.05 100 115.2 192 6.84 0.047 3 39.0
4 200 0 150 −150 1.0 0.043 0.75 0.43 4.05 180 180.0 300 12.37 0.047 1 15.0
5 600 0 450 −450 1.0 0.050 0.60 0.50 0.99 250 354.6 591 12.58 0.047 3 45.0;
**************************************************
Table BD(i, *) 'demands of each bus in MW'
Pd Qd
1 0 0
2 300 98.61
3 300 98.61
4 400 131.47
5 0 0 ;
**************************************************
Table LN(i, j, *) 'network technical characteristics' r x b limit
1 .2 0.00281 0.0281 0.00712 400
1 .4 0.00304 0.0304 0.00658 400
1 .5 0.00064 0.0064 0.03126 400
2 .3 0.00108 0.0108 0.01852 400
3 .4 0.00297 0.0297 0.00674 400
4 .5 0.00297 0.0297 0.00674 240;
**************************************************
LN(i, j, 'b') = 0;
LN(i, j, 'x')(LN(i, j, 'x') = 0) = LN(j, i, 'x');
LN(i, j, 'r')(LN(i, j, 'r') = 0) = LN(j, i, 'r');
LN(i, j, 'b')(LN(i, j, 'b') = 0) = LN(j, i, 'b');
LN(i, j, 'Limit')(LN(i, j, 'Limit') = 0) = LN(j, i, 'Limit');
LN(i, j, 'bij')LN(i, j, 'Limit') = 1/LN(i, j, 'x');
LN(i, j, 'z')LN(i, j, 'Limit') = sqrt(sqr(LN(i, j, 'x')) + sqr(LN(i, j, 'r')));
LN(j,i,'z')(LN(i, j, 'z') = 0) = LN(i, j, 'z');
LN(i, j, 'th')(LN(i, j, 'Limit') and LN(i, j, 'x') and LN(i, j, 'r')) = arctan(LN(i, j,
'x')/(LN(i, j, 'r')));
LN(i, j, 'th')(LN(i, j, 'Limit') and LN(i, j, 'x') and LN(i, j, 'r') = 0) = pi/2;
LN(i, j, 'th')(LN(i, j, 'Limit') and LN(i, j, 'r') and LN(i, j, 'x') = 0) = 0;
LN(j, i, 'th')LN(i, j, 'Limit') = LN(i, j, 'th');
Parameter cx(i, j);
cx(i, j)(LN(i, j, 'limit') and LN(j, i, 'limit')) = 1;
cx(i, j)(cx(j, i)) = 1;
Variable OL,OF, Pij(i, j), P(i), prow(i), qrow(i), K(i), P1, Q1, Qij(i, j), Pg(i), Qg(i),
Va(i), V(i), Qm1, Qmm1(i), Qmm2(i), Qrr1, Qm2, Qrr2, Qmw(i), Qres(i), Qmm(i);
Equation eq1, eq2, eq3, eq4, eq5, eq6, eq7, eq8, eq9, eq10, eq11, eq12, eq13, eq14,
eq15, eq16, eq17, eq18, eq19, eq20, eq21, eq22, eq23, eq24, eq25, eq26, eq27, eq28,
eq29, eq30, eq31, eq32, eq33;
```

```
eq1..
OL = e = sum(i, (Qg(i)*qrow(i))) + Qm1 + Qrr1;
eq2(i, j)cx(i, j)..
Pij(i, j) = e = (V(i)*V(i)*cos(LN(j, i, 'th')) – V(i)*V(j)*cos(Va(i) – Va(j) + LN(j, i,
'th'))) /LN(j,i,'z');
eq3(i, j)cx(i, j)..
Qij(i, j) = e = (V(i)*V(i)*sin(LN(j, i, 'th')) – V(i)*V(j)*sin(Va(i) – Va(j) + LN(j, i
,'th'))) /LN(j, i, 'z') – LN(j, i, 'b')*V(i)*V(i)/2;
eq4(i)..
Pg(i)GenD(i, 'Pmax') - BD(i, 'pd')/(Sbase) = e = sum(jcx(j, i), Pij(i, j));
eq5(i)..
Qg(i)GenD(i, 'Qmax') – BD(i, 'Qd')/(Sbase) = e = sum(jcx(j, i), Qij(i, j));
eq6(i)..
Qmw(i) = e = sum((j), (LN(i, j, 'x')*Pij(i, j)*Pij(i, j))/(V(j)*V(j)));
eq7(i)..
Qg(i)GenD(i, 'Qmax') = e = (GenD(i, 'Qmax'))/Sbase – (Qres(i));
eq8..
Qm1 = e = sum(i, ((Qmm(i)*GenD(i, 'a2'))+(GenD(i, 'a1')))*Qmm(i)) – (((Qmm
('1')*GenD('1', 'a2')) + (GenD('1', 'a1')))*Qmm('1')) + (((Qmm('1')*GenD('1',
'a2')) + (GenD('1', 'a1')))*Qmm('1')*2);
eq9..
Qrr1 = e = sum(i, (Qres(i)*Qres(i)*GenD(i, 'a2')))-(Qres('1')*Qres('1')*GenD('1',
'a2')) +(Qres('1')*Qres('1')*GenD('1', 'a2'));
eq10(i)..
abs(V(i) – 1)/1 = l = 0.01;
eq11(i)..
((1) – (V(i)*V(i)) + 1)*sum(j, (LN(i, j, 'bij'))) = g = Qmm(i);
eq12..
P1 = e = sum(i, Pg(i)*prow(i)) ;
eq13..
Q1 = e = sum(i, Qg(i)*qrow(i)) ;
eq14(i)..
Qmm(i) = g = Qmw(i);
eq15(i)..
Pg(i) = l = GenD(i, 'Pmax')/Sbase;
eq16(i)..
Pg(i) = g = GenD(i, 'Pmin')/Sbase;
eq17..
OF = e = sum(i, (qrow(i)*Qg(i)) – (GenD(i, 'a2')*(Qg(i) – (GenD(i, 'Qbb')*0.5/
Sbase))*(Qg(i) – (GenD(i, 'Qbb')*0.5/Sbase))) – (GenD(i, 'a1')*(Qg(i) – (GenD(i,
'Qbb')*0.5/Sbase))) – GenD(i, 'a0'));
eq18(i)..
Qg(i) = g = GenD(i, 'Qmin')/Sbase;
eq19(i)..
Qg(i) = l = GenD(i, 'Qmax')/Sbase;
```

```
Model loadflow/eq1, eq2, eq3, eq4, eq5, eq6, eq7, eq8, eq9, eq10, eq11, eq12,
eq13, eq14, eq15, eq16, eq17, eq18, eq19/;
Va.up(i) = pi/2;
Va.lo(i) = −pi/2;
Va.l(i) = 0;
Va.fx(slack) = 0;
Pij.up(i, j)((cx(i,j))) = 1*LN(i, j, 'Limit')/Sbase;
Pij.lo(i, j)((cx(i, j))) = −1*LN(i, j, 'Limit')/Sbase;
Qij.up(i, j)((cx(i, j))) = 1*LN(i, j, 'Limit')/Sbase;
Qij.lo(i, j)((cx(i, j))) = −1*LN(i, j, 'Limit')/Sbase;
V.lo(i) = 0.95;
V.up(i) = 1.05;
V.l(i) = 1;
onecho > " bilevel qrow
max OF Qg eq17 eq18 eq19
offecho
solve loadflow use emp min OL;
```

Appendix B
Reactive Power for Voltage Support

Consider two buses i and j in an N bus system (Fig. B.1).

The reactive power generation (Q_i) required at ith bus when there is no reactive power generation at jth bus is given by

$$Q_i = \frac{V_i^2 - V_j^2}{X_{ij}} \tag{B.1}$$

The minimum reactive power ($Q_{vi}^{\min}$) required for shipment of active power at ith bus (Wang and Xu 2004) is given by

$$Q_{vi}^{\min} = \sum_{j \in N} \frac{X_{ij}(P_{ij})^2}{V_j^2} \tag{B.2}$$

where
P_{ij} is the active power flow between ith bus and jth bus
X_{ij} is the (i,j)th element of bus reactance matrix
V_j is the voltage at jth bus
Equation B.2 is considered as the lower bound for reactive power required for voltage support. Let Vref_i be the desired voltage at ith bus. Then reactive power required to maintain desired voltage at ith bus is

$$\text{Qref}_i = \frac{\text{Vref}_i^2 - \text{Vref}_j^2}{X_{ij}} \tag{B.3}$$

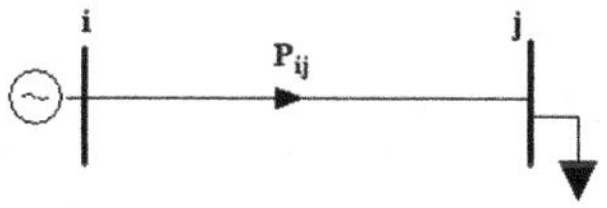

Fig. B.1 Generator-load bus system

D. Jay and K. Shanti Swarup, *Market Operation for Reactive Power Ancillary Service*, Springer Tracts in Electrical and Electronics Engineering,
https://doi.org/10.1007/978-981-99-6952-4

Thus the additional reactive power generation required at ith bus, if all buses connected to it are load bus with no reactive power support is given by

$$\Delta Q_i = \sum_{j \in N} \left[\frac{\mathrm{Vref}_i^2 - V_i^2}{X_{ij}} + \frac{V_j^2 - \mathrm{Vref}_j^2}{X_{ij}} \right] \tag{B.4}$$

However to determine the actual reactive power generation required for voltage support alone at the bus in real time is challenging. Hence in this work an upper bound is set by considering the worst case for value of $(V_j^2 - \mathrm{Vref}_j^2)$. Thus upper bound for reactive power generation ($Q_{vi}^{\max}$) required at ith bus is set as

$$Q_{vi}^{\max} = \sum_{j \in N} \frac{\mathrm{Vref}_i^2 - V_i^2 + 1}{X_{ij}} \tag{B.5}$$

Thus reactive power generation required at ith bus in a localised market area, for maintaining bus voltages, defined as Q_{vi} is a decision variable for ISO model with lower and upper limits set by $Q_{vi}^{\min}$ and $Q_{vi}^{\max}$ respectively.

Reference

Wang Y, Xu W (2004) An investigation on the reactive power support service needs of power producers. IEEE Trans Power Syst 19(1):586–593

The manufacturer's authorised representative in the EU is Springer Nature Customer Service Centre GmbH, Europaplatz 3, 69115 Heidelberg, Germany. If you have any concerns regarding our products, please contact ProductSafety@springernature.com

Printed and bound by CPI Group (UK) Ltd, Croydon, CR0 4YY
01/10/2026
02225877-0006